A Stake in the Future

Mary Louise McAllister
Cynthia Jacqueline Alexander

A Stake in the Future: Redefining the Canadian Mineral Industry

UBCPress / Vancouver

Published in association with the Centre for Resource Studies, Queen's University at Kingston

Printed in Canada on acid-free paper ∞

ISBN 0-7748-0603-6 (hardcover)
ISBN 0-7748-0602-8 (paperback)

Canadian Cataloguing in Publication Data

McAllister, Mary Louise, 1957-
A stake in the future

Includes bibliographical references and index.
"Published in association with the Centre for Resource Studies, Queen's University at Kingston."
ISBN 0-7748-0603-6 (bound)
ISBN 0-7748-0602-8 (pbk.)

1. Mineral industries – Government policy – Canada. 2. Mineral industries – Social aspects – Canada. 3. Whitehorse Mining Initiative (Canada). I. Alexander, Cynthia Jacqueline, 1960- II. Queen's University (Kingston, Ont.). Centre for Resource Studies. III. Title.

HD9506.C22M222 1997 338.2'0971 C97-910088-7

UBC Press gratefully acknowledges the ongoing support to its publishing program from the Canada Council, the Province of British Columbia Cultural Services Branch, and the Department of Communications of the Government of Canada.

Set in Stone by Val Speidel
Printed and bound in Canada by Friesens
Proofreader: Rachelle Kanefsky
Indexer: Annette Lorek

UBC Press
University of British Columbia
6344 Memorial Road
Vancouver, BC V6T 1Z2
(604) 822-5959
Fax: 1-800-668-0821
E-mail: orders@ubcpress.ubc.ca
http://www.ubcpress.ubc.ca

Contents

Tables and Figures

Tables

Figures

Acknowledgments

This book, much like the national Whitehorse Mining Initiative (WMI), required the perspectives, insights, and assistance of a great number of people. George Hood, former Director of the Centre for Resource Studies (current Associate Vice-Principal, Research, Queen's University), and Michael Doggett, Interim Director of the Centre for Resource Studies, Queen's University, provided encouragement, direction, and enthusiasm at all the right times along the way. Their steadfast faith in the project and the authors was very much appreciated. We also thank CRS for its financial support. The comments provided by those who served as peer reviewers for the manuscript were very helpful and constructive. We did our best to respond to their suggestions and thank them for their suggestions.

George Miller, President of the Mining Association of Canada, has garnered tremendous respect within all the communities of interest involved in the Whitehorse Mining Initiative. Sustainable mineral development in the next century will depend on the leadership of individuals such as Miller who know what needs to be done and have the courage to do it. He took time from his busy schedule to provide us with valuable insights into the WMI process and the workings of the mineral industry. Dan Johnston, a facilitator of the WMI and the BC Advisory Council on Mining, acted as an indispensable bridge in the WMI, ensuring that all participants continued to communicate. During our interviews, Johnston was frequently cited as one of the most important individuals in ensuring the successful signing of the accord. Dan patiently explained to us the complexities involved in the art of negotiation and facilitation.

Participants in the Whitehorse Mining Initiative and the BC Advisory Council on Mining kindly responded to all our questions during a long interview process and numerous follow-up phone calls, and read over relevant chapters. Bruce McRae, then A/Deputy Minister of the BC Ministry of Energy, Mines and Petroleum Resources,[1] read sections of the manuscript, providing a helpful critical perspective and objective advice. The following list includes the names of many people who took time to send information and to be interviewed during the past year: Tony Andrews (Managing Director, PDAC), Jerry Asp (Consultant), Art Ball (Director of Mines, Manitoba Department of Energy and Mines), Don Barnett (Assistant Deputy Minister of Mines and Energy, NB), Richard Boyce (President, Local Union 7619, USWA), Keith Conn (Coordinator, Environment, Assembly of First Nations), Paul Dean (ADM, Mineral Resource Management Branch, Department of Mines and Energy, NF), the Hon. Don Downes (Minister of Natural Resources, NS), the Hon. Anne Edwards (then BC Minister of Energy, Mines and Petroleum Resources), John B. Gammon (ADM, Mines and Minerals Division, Ministry of Northern Development and Mines, ON), David Hopper (Land-Use Planner, Department of Natural Resources, NB), Lois Hooge (Head, Whitehorse Mining Initiative Secretariat), Douglas Hyde, (Canadian Environmental Network), Ed Huebert (VP, Manitoba Association of Mining), Bob Keyes (VP, Economic Affairs, Mining Association of Canada), Gary Livingstone (President and CEO, Mining Association of British Columbia), Hans Matthews (President, Canadian Aboriginal Minerals Association), Hugh MacKenzie (Assistant to the National Director, USWA), Ed Mankelow (BC Wildlife Federation), Dan McFadyen (ADM, Resource Policy and Economics, Saskatchewan Energy and Mines), Graeme McLaren (Manager, Land-Use Policy, EMPR), Gary MacEwen (Natural Resources and Energy, NB), Bruce McKnight (VP, Corporate Affairs, Westmin Resources), Irene Novaczek (Chairperson, Ocean-Caucus-PEI Environmental Network), Brian Parrott (then A/ADM, BC Energy, Mines and Petroleum Resources), George Patterson (Director, Mineral Policy, Energy, Mines and Petroleum Resources, Government of the NWT), Jack Patterson (Managing Director, BC and Yukon Chamber of Mines), Bert Pereboom (Senior Programme Officer, Research Department, USWA), Pat Phelan (Executive Director, Mines and Minerals, Department of Natural Resources, NS), Heather Robertson (Senior Planner, Ontario Ministry of Northern Development and Mines), Walter Segsworth (President, Westmin Resources), Peter Smith (Staff Representative, Canadian Auto Workers), Darlene Smith (Department of Fisheries and Oceans), Bill Toms (Chief, Business Resource Tax Analysis, Department of Finance), Robert Van Dijken (President, Yukon Conservation Society), and Alan Young (Canadian Environmental Network).

Other individuals, although not participants in the Whitehorse Mining Initiative, also gave freely of their time and expertise. They include Dan Adamson of BC Parks (who patiently explained the province's new land-use planning processes), Carylin Behn (CRG Projects), who carefully read the sections on Aboriginal perspectives and provided a thoughtful and very helpful critique, Annie Booth (Environmental Studies, UNBC), Greg Prycz (Political Science, Acadia University), Beverly K. Therrien (Edmonton, AB), and Paul Banfield and his colleagues at Queen's University archives.

Don Manson (of the political science program at UNBC) provided invaluable assistance in helping to pull the first draft of the manuscript together. He chased down sources and edited, reviewed, and formatted text with expertise, considerable forbearance, and good humour. Cheerful and professional secretarial support was provided by Maggie Clarke, Lois Crowell, Bev Schroeder (UNBC), Donna Stover (CRS, Queen's), and Leanne Wells (Acadia University). Thanks also to members of our departments for their collegial support.

We have tried to be accurate and objective in our portrayal of the many different perspectives as they relate to mining and the WMI. None of the individuals mentioned here should in any way be held responsible for possible misinterpretation or errors of fact. They are the authors' own.

Similar to the Whitehorse Mining Initiative, the scope of this project was of a magnitude not originally anticipated. During the WMI process itself, much was expected of participants in the way of concessions, compromise, and conciliation. This was also the case with our families, who were tremendously supportive during the whole project, and it is to them that this work is dedicated.

Introduction

> Our vision is of a socially, economically and environmentally sustainable, and prosperous mining industry, underpinned by political and community consensus ... The *Principles and Goals* that we have adopted represent a major and historic first step toward revitalizing mining in Canada. They point to changes that can restore the industry's ability to attract investment for exploration and development and, at the same time, ensure that the goals of Aboriginal peoples, the environmental community, labour, and governments will be met.[1]

In the contemporary policy world, decision-makers are attempting to foster a new spirit of cooperation in the management of natural resources. Governments are mired in jurisdictional conflicts, struggling with massive debts and public cynicism, and are confronted with warnings of impending doom to the economy and environment. Moreover, with rising levels of education, improved access to information, and the growth of a 'rights' oriented society, there is increasing pressure for direct public participation in the decision-making process. As a result, recent years have witnessed a proliferation of roundtables and public consultation processes at all jurisdictional levels.

The Whitehorse Mining Initiative (WMI) was one such effort. This was a national initiative that received its name at an annual meeting of the mines ministers in 1992. The WMI was a process that would encompass all of Canada. It was an effort by the mineral industry to forge new alliances with other groups who had a stake in the resources affected by mineral development. The mining industry has served as a traditional pillar of the Canadian economy, yet it no longer engenders the same level of support from the general public and the government that it once commanded. The creators of the WMI recognized that consensus-building and integrated resource management is now a widespread policy approach to which they must adapt if mining is to continue to thrive in Canada.

The final accord represents a considerable achievement in the history of mining. Because of its scope and unique approach, the WMI has attracted international attention among stakeholders in many mining countries. As in the case of Canada, these countries are looking to develop new ways of fostering a policy environment that accommodates the concerns of a plurality of interests while allowing for sustainable mining. The WMI represents a first important step toward that goal.

The animating thesis of this text suggests that a consensus-based decision-making approach with all its inherent promises and problems has now become an inevitable component of resource policy. The implications of such an approach, therefore, require careful study. Extensive participation by interest groups and others in resource decision-making raises many concerns about democratic representation, as well as possible adverse impacts on policy effectiveness and efficiency. Nevertheless, there appear to be few viable alternatives in a policy-making world where vocal members of society continue to assert their rights to participate and help formulate public decisions. As the Canadian experience so lucidly demonstrates, the politics of big business, big government, and executive federalism is no longer acceptable to many competing communities of interest. More than ever before we have moved into the world of compromise and conciliation. Canadians have long been familiar with such negotiations, but support inevitably diminishes if the outcomes appear unpromising or inconclusive. Consensus-based approaches need to offer some results if they are to be successful.

The Whitehorse Mining Initiative was unique, in part because it was initiated by industry – a radical step for a very conservative sector. Moreover, its proponents were ambitious and bold. Those individuals within the mining industry needed to be able to persuade other members of the industry that the WMI was worth their time. Their personal credibility within the industry itself and in the eyes of a sceptical public was, in many ways, on the line. Furthermore, the promoters of the WMI had to persuade the rest of the mining industry to support something beyond their usual terms of reference. This time, members of the industry were not asked to explore something as tangible as a promising ore body. They were asked instead to participate in the very unfamiliar terrain of consensus-building, democratic communications, and policy-making.

An examination of the WMI allows us to explore a number of issues on different levels of analysis. On the first level, we examine the industry itself, including its role in Canada's economy and society, the current challenges it faces, and the policy issues that need to be addressed. On the second level, we observe increasing pressures by vocal members of the public to be consulted and included in the policy-making process. This is particularly evident in issues relating to natural resource policy. The WMI represents an attempt to accommodate those demands and improve public acceptance of both industry and the policy decisions made surrounding mineral exploration and development. On a third level, we examine whether multi-stakeholder initiatives such as the WMI and other consultative processes can be sustained and effectively implemented. Finally, we consider the impact of this new policy style on our democratic institutions. Are these new approaches, for example, simply providing for the dominance of special interests – albeit a wider array of them than was previously the case?

In the following chapters, a series of issues underpinning the WMI are addressed. In the first chapter, we discuss the forces that led up to the Whitehorse Mining Initiative and consider how well this consultation process fits within Canadian decision-making processes. Increasingly, these processes are being used to inform land-use decisions and help determine the public interest. Such an approach is not without its difficulties in a liberal democratic state where resource values conflict and issues concerning property rights, democratic representation, and political accountability have yet to be resolved. The issues dealt with in this chapter are being broached in most areas of land-use decision-making today as people wrestle with the notion of sustainable development.

Chapter 2 examines some of the competitive difficulties of the mineral industry and the challenges to its future viability. These challenges are considered to be a result of international competitiveness as well as an industry perception that the regulatory environment in Canada is no longer hospitable to exploration and development activities. Part of the difficulty is that there is not a great deal written about the mineral industry and it does not achieve much salience on the public agenda. These competitive concerns stimulated the creation of the WMI.

Chapter 3 explores the competing ideological values and interests of the different groups that gathered together at the WMI. Each of the policy communities themselves are disparate and, therefore, do not easily fit into tidy categories for analysis. Nevertheless, they share some general world views that are explored in this chapter. In addition, there is a brief description of what some members of these groups hoped to achieve in the WMI.

Chapter 4 sets out the structure and the processes of the WMI. It is assessed within the context of a discussion about the variables that must be considered when devising successful consultation processes.

Chapter 5 examines the issues, recommendations, and goals of the issue groups, leadership council, and the other participants in the WMI that led to the writing and signing of a final accord. Here many issues and concerns are addressed, including the investment environment, mine reclamation, protected areas, land access, Aboriginal concerns, health and safety, and others. Given the breadth of the issues discussed, the fact that a consensus document was signed was a remarkable accomplishment. The accord and the reports of the issue groups can serve as a blueprint for future mineral policy or for other countries that are exploring a similar approach to sustainable mining.

Chapter 6, the most comprehensive section, explores the actual implementation of the accord in various jurisdictions. It is often at the administration and implementation stage that such initiatives lose momentum. This appears also to be the case with the WMI. Despite this, progress has been made in many areas. An examination of some of these new programs and policies serve as useful examples of how the principles of the initiative can be applied.

Chapter 7 considers whether or not the WMI was a success from the perspectives of the various policy communities that participated in the initiative. Participants met a little over one year after the accord was signed and offered comments and criticisms about the efforts undertaken in the first year. They also made concrete suggestions for accomplishing some of the goals established during the WMI process.

Chapter 8 concludes the book with a series of recommendations for future directions in the implementation of the WMI. This consensus-based approach could serve as a very useful model to resolve difficult land- and resource-use policy issues. It is a model that could be applied from the community level to international arenas. Furthermore, the lessons offered by the Whitehorse Mining Initiative are applicable to other resource-related policy areas including forestry, land use, oil and gas development, and fisheries. Immediate and concrete benefits were generated by the WMI process. In particular, it has served to develop a broadly accepted framework for sustainable mineral activity in Canada, fostered important communications networks, and encouraged a more cooperative policy environment. As is to be expected, it will be a challenge to maintain momentum in the implementation of the recommendations. The creation of the WMI and the signing of an accord demonstrates that much can be achieved. An examination of the initiative has also generated some more fundamental, intriguing questions. These questions include the implications of consultation and multi-partite approaches for public decision-making, administrative reform, and liberal democracy. Those involved in exploring the prospects for sustainable development through public consultation, roundtables, and democratic governance are, in many ways, forging through rough terrain. Nevertheless, the WMI does provide some direction for those attempting to achieve a consensus in resource policy-making.

Abbreviations

ACM	Advisory Council on Mining (BC)
ADM	Assistant Deputy Ministers
AFN	Assembly of First Nations
CAMA	Canadian Aboriginal Minerals Association
CANMET	Canada Centre for Mineral and Energy Technology
CAW	Canadian Auto Workers
CEA	Canadian Environmental Assessment Act
CH	Canadian Heritage-Parks
CIM	Canadian Institute of Mining, Metallurgy and Petroleum
CLURE	Commission on Land Use and the Rural Environment (NB)
CORE	Commission on Resources and Environment (BC)
DIAND	Department of Indian Affairs and Northern Development
EARP	Environmental Assessment and Review Process
EC	Environment Canada
EIR	Environmental Impact Statements
EMR	Energy, Mines and Resources (Canada)
LDC	Long-Distance Commuting
LRMP	Land and Resource Management Plans
MAC	Mining Association of Canada
MAM	Manitoba Association of Mining
MDA	Mineral Development Agreement
MDAP	Mine Development Assessment Process
MDC	Mine Development Certificate
MEMPR	Ministry of Energy, Mines and Petroleum Resources (BC)
MITAC	Mining Industry Training and Adjustment Council
MNDM	Ministry of Northern Development and Mines (ON)
MPS	Mineral Policy Sector (Canada)
NRCan	Natural Resources Canada
OMNR	Ontario Ministry of Natural Resources
PAS	Protected Areas Strategy

PDAC	Prospectors and Developers Association of Canada
PPDA	Porcupine Prospectors and Developers Association
SEM	Saskatchewan Energy and Mines
USWA	United Steelworkers of America
WMI	Whitehorse Mining Initiative
WWF	World Wildlife Fund

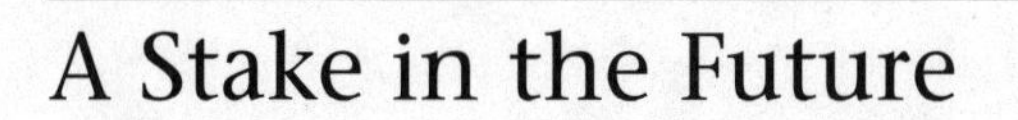

A Stake in the Future

1
Surveying the Terrain

The Whitehorse Mining Initiative (WMI)

In February 1993, a large group assembled in a conference room in downtown Toronto for the first meeting of the Whitehorse Mining Initiative (WMI). Gathered together were people with different world views, ideological perspectives, and interests. They included leaders of industry and labour, deputy ministers of mines, well-known members of Aboriginal communities, and politically astute environmentalists. Participants were to consider a new way of dealing with the country's natural resources – one that would recognize a diversity of values, thereby establishing a long-term plan for the sustainable development of mineral resources. The WMI and its succeeding initiatives are attempts to adapt to the changing policy world in ways that might allow the mineral industry to remain competitive.

The genesis of the WMI can be traced back to September 1992 when the Mining Association of Canada presented a unique proposal at the Canadian mines ministers' conference in Whitehorse, Yukon. The proposal was for a 'Whitehorse Charter' that might help combat structural problems that were affecting the competitiveness and long-term sustainability of the Canadian mineral industry. The proposal called for the formation of a comprehensive plan of action agreed to by influential policy communities: government, industry, labour, Aboriginal, and environmental groups. The Whitehorse Mining Initiative was named after the city where the new plan was announced.

The WMI represents an unprecedented effort initiated by the mineral community to bring diverse communities of interest together. As such, it has adopted an ambitious policy design, a 'radically new approach' that calls for dealing strategically with structural problems 'rather than dealing with individual problems in a piecemeal fashion.'[1] The plan was the creation of the Mining Association of Canada (MAC), spearheaded by its president, George Miller. Miller and his colleagues at MAC had to persuade the members of various mining-related associations in Canada (many of them very sceptical) that such an effort would be worth their time. Once that task was accomplished, governments and representatives from non-governmental organizations and labour and Aboriginal groups would also need to be convinced that the industry was serious in pursuing such a collaborative effort. The initiative did not seem to be in keeping with the individualist, classical liberal ideology of the mineral industry. Yet Miller commands considerable respect from those who know him, and several leaders in the mineral industry could see the value in what was being attempted and were persuaded to support the project.

At the February 1993 multi-stakeholder meeting, participants agreed that the outcomes of the WMI should include:

- a strategic vision for the minerals and metals sector
- a series of 'accords' or 'partnership agreements' between/among various stakeholders and industry
- options and recommendations for policy or regulatory changes consistent with priorities in each jurisdiction
- and/or the establishment of new consultative mechanisms to ensure the continuation of communication and cooperation, and to facilitate the management and/or resolution of conflicts.[2]

An accord was signed in the fall of 1994 at the annual meeting of mines ministers in Victoria, BC. The WMI reflected a decision-making approach that is increasingly becoming an expected practice in policy-making in Canada and elsewhere. This new approach employs a 'multi-stakeholder' consultation process to accommodate and to integrate competing public perspectives about the appropriate management of natural resources.

A New Policy Approach – Consultation Processes

Some observers may wish to characterize this new policy style as a nebulously defined process-driven approach, informed by an indeterminate number of interested participants attempting to work together toward a vaguely defined goal of sustainable development. Consensus-based processes do raise many troubling questions, including whether or not the participants are appropriate representatives of the public interest and whether the results of these processes lead to the best policy decision. That said, however, it could also be argued that such new processes represent 'an effort to expand the traditional bargaining system to include both labour and environmentalists as well as business and government. The objective [is] consensus.'[3] Extensive public consultation – although a lengthy and often frustrating process – could result in a better and more effective policy decision because the policy has been carefully scrutinized from a variety of angles before implementation. Equally important, decisions based on these 'roundtables'[4] carry with them a legitimacy that governments need to maintain public support.

In Canada, roundtables are responses to a number of developments that affect policy-making, including the evolution of a 'rights' oriented society,[5] a diverse set of pressure points, the complex federal-provincial environment, public disenchantment with government and its increasing lack of legitimacy, and the need for industry to share government agendas with other competing interests. George Hoberg suggests that the concept of multi-stakeholder consultations emerged in the early 1980s when Environment Canada contacted the Niagara Institute to develop a consultative approach that could be used to resolve environmental conflicts:

> While there is no necessary logical or causal relation between the ideas of sustainable development and multi-stakeholder consultations, there appears to be a strong elective affinity between the two concepts among core actors in the environmental policy community. Both are based on the idea that corporate interests in development can somehow be reconciled with interests in environmental protection. Political conflict over unavoidable trade-offs were perceived as unproductive, and consensus could be achieved only if a proper process was designed to allow the competing stakeholders an opportunity to communicate with each other.[6]

As in the case of other natural resources, the development and regulation of minerals must reflect and provide for the diverse interests that, increasingly and ever more vocally, have been demanding their right to participate in the policy process. Many leaders of the mining community are now recognizing that new collaborative, consensual policy strategies are necessary if the industry is to survive, let alone thrive. The alternative policy process, that of adversarial politics, encourages opposition, hinders investment, and inhibits long-range planning of sustainable resource use. The opportunity and desire for compromise becomes paralyzed and participants get swept up in long, drawn out court battles.

The polycentric nature of the problems facing the mining industry – with numerous issues intricately intertwined – required the adoption of an integrative resource management approach. Among the issues facing industry are human resource development, taxes and charges, environmental assessment and permitting, the environmental regulatory process, land access, land claims, and Aboriginal self-governance. In addition, public perceptions about the industry are generally misinformed and outdated. Although there is little research on the subject, there are also indications that the mineral industry may be poorly regarded, not only by environmental activists and Aboriginal groups, but by the general public as well. Public perceptions influence government agendas. The Whitehorse Mining Initiative is part of the effort to address these concerns.

Digging into the Past

The Canadian mineral industry suffered a significant decline since the early 1980s. Mineral profits, investment, and reserves went into a downward spiral. Throughout the 1980s and early 1990s, mine closures have been increasing and openings decreasing. By the middle of the 1990s, the picture had improved. In 1994, there were more mine openings than closings. By 1995 high commodity prices and international demand led to a much stronger economic performance. Many existing mines, however, are nearing the end of their productive lives. Furthermore, competing demands for the land and growing global competition are forcing the Canadian industry to either rethink its traditional approaches to mining, or continue to watch the erosion of its position as a world leader in mineral exports.

There have been many explanations for the difficulty in which the industry finds itself. Mineral economists, while rejecting traditional arguments of resource scarcity, nevertheless acknowledge that the known, easily accessible economic deposits are becoming depleted in Canada. This means that industry and governments need to devise a longer-term strategy in order to explore for, develop, and produce minerals. Developing countries are able to attract capital because they are perceived to be underexplored. If this is the case, then accessible, high-grade deposits should be easier and cheaper to find. Production costs are generally lower in these countries because labour is cheaper and regulations are more favourable to mineral development.[7] In the years following the Second World War, there was an antipathy to foreign direct investment and the associated higher costs involved in production. As a result, the majority of the Western world's mineral exploration (over 70 per cent) took place in Canada, the United States, and Australia. Now, many developing countries, with their promising resources, are actively pursuing foreign investment and are in a position of being able to reduce their former cost and productivity disadvantages. In short, Canada, Australia, and the United States will no longer be able to retain such a commanding lead.[8]

Domestically, it is a very different world that the mineral industry is now encountering – a contrast to the fairly recent past. Throughout Canada's history, the public interest was often defined as synonymous with economic development. The national political agenda was to build a country with the wealth generated by its rich natural resources. The role of governments was to ensure that economies were thriving and business was booming. Western liberal societies were caught up in the drama and the promise of new technologies, rapid industrialization, and an expanding population.

In Canada, a close, interdependent relationship between government and business was notable in the resource development policies of the time.

Mineral resource policies were supportive of resource development and industry encountered relatively few barriers in its quest to explore for, and extract, minerals. As Marsha Chandler states:

> The overriding goal – to facilitate the exploitation of the resources – has always been considered essential to economic development. The exploitation and the development of resources have for the most part been left to the private sector, and the role of provincial governments has been to stimulate private activity through tax incentives, provision of ancillary services, infrastructure, and direct subsidies.[9]

The devastating impact of the Great Depression of the 1930s made it clear that individuals could no longer be seen as solely responsible for their economic well-being. Government's redistributive activities became justified on both social and economic grounds. Keynesian economics suggested that government intervention in the economy could moderate dramatic swings in the business cycles.

The Canadian state developed in a way that fostered the nascent resource industries. As S.D. Clark points out, 'Canada has been what the late H.A. Innis called a "hard frontier." The exploitation of her resources has required large accumulations of capital, corporate forms of business enterprise, and state support.'[10] It was also state activity that redistributed the wealth generated through the activities of the primary industries to other sectors or interests. The 1920s saw tremendous expansion in base-metal mining. In the 1930s, when other sectors were suffering through the depression, gold mining continued to expand. The 'frontier' mentality of prospectors and developers fuelled their resentment toward governments' economic activities. According to H.V. Nelles:

> An ardent individualism provided intellectual reinforcement for the industry's economic resentment of government interventionism. The industrial symbol was the solitary prospector who made his way upward in the world through sheer wit and pluck without assistance from any quarter. That the symbol did not correspond to a reality of financial intrigue and technological intricacy attested not so much to its irrationality as to its strength and the psychological need it fulfilled.[11]

It should be noted that such a characterization need not be confined to the mineral industry nor need it be historically bound. E.P. Herring made an enduring observation sixty years ago that groups will seek government assistance and special consideration while at the same time resenting state 'encroachment' into 'private' affairs.[12]

Resource development was consistent with many of the national and province building goals throughout much of Canada. As such, resource development was often viewed by governments as a way to promote the public interest. While the methods and means of achieving the development might be contested, the ultimate objective was not. In a historic discussion of British Columbia's resource politics in the 1960s, E.R. Black suggested that the purpose of natural resource exploitation was not the subject of dispute. The issues were altogether different:

> When conflicts do arise, they almost inevitably centre on the means and timing of exploitation, and not on its objectives. Even when the perspective adopted is that the resources belong to the people and ought to be developed for their benefits, the questions, Which people? and Today or tomorrow? remain unasked and unanswered. Should resources be developed to maximize both the immediate revenue return to the provincial treasury and the economy in terms of the gross provincial products? Some would say, 'Yes!' Others might claim there is always a fundamental obligation to consider alternative uses of the resources and their implications ...[13]

In the latter part of the twentieth century, however, such questions are now being asked in British Columbia and elsewhere. Public values and attitudes toward resource development are changing in response to shifting demographic patterns, new recreational interests, and transforming political, economic, and physical environments.[14] The interests of resource developers conflict with those of other groups. The modern state provides many indirect and direct communication channels through which competing interests can articulate their concerns. These include the courts, the legislatures, the uncoordinated and disjointed government apparatus, national and sub-national government, and the media. The mining industry no longer enjoys the government and general public support that it received until recently. Furthermore, the state itself lacks a framework that is supportive of coordinated development initiatives. One of the reasons for this is the growing complexity of the modern state that limits the ability of governments to provide a consistent, predictable regulatory environment that would attract investment capital to resource-based projects. A plurality of conflicting demands on the Canadian federal state and escalating land-use disputes contribute to investment uncertainty.

In response to competitive threats, the mineral industry has generally attempted to find a technological fix or a market-oriented solution to its difficulties. With growing concerns about the global environment, increasing overall global consumption of goods, and the competing demands of society, such fixes are not enough. The biggest variable in business competitiveness is the political arena. George Miller provides this analysis:

> Hugh O'Driscoll began his speech to the recent Mines Ministers' Conference in Victoria by referring to a cartoon character called Ziggy. Ziggy is driving along a highway. In the first frame he sees a sign: OLD ERA ENDING. In the second frame a sign: NEW ERA BEGINNING. In the third frame: PREPARE TO PAY THE TOLL.
>
> According to O'Driscoll, the passing era is represented by a mechanistic view of the world, in which science gives all the answers, and the best solution or policy is based on technical analysis. The new era is characterized by an organic view of the world, in which everything is connected to everything else. In the new era, decisions are not reached through analysis carried out by technocrats, but are rather worked out in society through the clash of values and interests.[15]

It has been 100 years since the 'scientific management' approach was first ardently pursued by many as a rational and efficient way to achieve public goals. Twentieth-century governance has often been informed by a belief that rationality and science can lead to effective decisions informed by objectively gathered facts. When decisions are based on scientific expertise, however, there is little room for participation by interested members of the public who are unable to communicate in the language of the specialist. Traditional approaches, particularly when they concern land-use and environmental issues, are increasingly viewed as serving a narrow interest.

While scientific analysis still plays an important role in the policy-making arena, the traditional role of science is now being reconsidered. This is particularly the case in a multi-stakeholder setting, in which environmentalists sit next to leaders of industry who sit beside Aboriginal representatives. John Dryzek claims that it is through reasoned discourse that the privilege which has been given to experts can be reclaimed.[16] It has also been argued that the task is not to 'supplant the knowledge supplied by experts and interest group leaders,' but to design a consultation process which incorporates a fuller spectrum of views.[17]

This is not an easy task. There are no clear-cut solutions to the complex, interdependent policy arena where interested participants embrace competing ideological perspectives. As Geraldine DeSanctis notes:

> There are competing perspectives on the nature and causes of the dilemmas and no right or wrong answers. Alternative solutions may be identifiable, but how to analyze or choose among them, and how to implement solutions once decided, are less clear-cut. Further, decisions about how to address the dilemmas can have substantial implications for the parties involved, as well as for the larger community or society. How should wicked environmental dilemmas be conceptualized? How can solutions be developed, analyzed, and implemented?[18]

The 'public interest' is now being determined by an ever-broadening constituency of competing stakeholders. Interpreting that public interest and ascertaining how the mineral industry fits into that definition is an issue that governments must work out within a complex arena of conflicting public values, pressure points, and political influence.

Defining the Public Interest in a Liberal-Democratic State

Despite the current pressure for public participation in decision-making, it is worth noting that interest group politics is not an invention of the past few decades. Those groups that seek to influence policy have been around for a long time. In Canada, lobbying has been around since Confederation and the railways. To be sure, the participants may have been fewer in the past. As E.P. Herring's 1930s essay noted, however, as the state undertakes a growing number of services, different interests increasingly develop a 'stake' in government activities: 'The actual point at which their right to pursue freely their own aims comes into conflict with the fulfilment of the broader aims of the community is a matter for judgment as particular cases arise. But the civil servant has no clear standard upon which to act. In a word, our present bureaucracy has accumulated vast authority but it lacks direction and coordination.'[19]

The difficulty, as indicated by Herring, is twofold. First, the administrative machinery of the growing state is not a cohesive whole that can respond in a unified way to the general public interest. It is fragmented into a loose association of national and sub-national governments that are further compartmentalized into departments, regulatory authorities, and agencies – all with their different and competing constituencies that must be served. Second, the public interest itself is ill-defined in the liberal democratic state. The public interest is often defined in the context of a particular policy process. In the process of soliciting public responses to a policy initiative, governments will receive a collection of individual reactions or responses. When these responses are arranged around a particular issue, group negotiations may occur. As diverse groups vie for the attention of the government, the undifferentiated mass public and its 'interest' becomes relatively unimportant in the policy context.[20]

In an effort to achieve some policy influence, both public and private interest groups often attempt to influence public policy by suggesting that their goals will promote the public interest. They do so because the criterion and justification for government action is based on some notion of the public interest.[21] William D. Coleman points out that business activities are often characterized as being associated with a more 'general' interest because such activities will generate jobs and growth for the economy. On the other hand, groups that are not associated with economic growth are more often characterized as 'special' interests.[22] Interestingly enough, the opposite argument has also been made; that is, economic interests are often considered to be 'special' interests. Business groups are often seen as more self-serving and 'narrow' because of their pecuniary interests than those groups who seek collective and social benefits.[23] From a government perspective, therefore, it is difficult to know who should be included in a multi-stakeholder approach to policy-making to ensure a general public interest is represented. Even if the public interest is often heavily influenced by the bargaining process between various players (as some observers might conclude), a general justification for such decision-making in Western societies is ultimately informed by the values of a liberal-democratic system. In liberal democracies, recognition of both private and public interests are necessary to the continued functioning of the existing state system.

Liberal Democracy and Competing Resource Values

The liberal-democratic system of values recognizes the rights of individuals to possess certain freedoms, to own property, and to participate in the political process. T.C. Pocklington suggests that notable among countries usually recognized as liberal democracies 'is the pre-eminence in their economies of privately owned and operated economic enterprises.'[24] Economic growth has often been considered a precondition for democracy. In Canada, the abundance of natural resources has contributed to a very high standard of living and, some would argue, a moderate political climate.[25] According to one observer: '[Governments] must satisfy an electorate while ensuring the continued economic viability of the system. They must recognize popular values and deal with the needs and demands of various interest groups. And they must act, appear to act, and satisfy themselves that they are acting, to the best of their ability, in the public interest.'[26] In the area of natural resources, the public interest is not easily defined. Who benefits from the exploitation of the natural resources? Who should have access to those natural resources? Disagreement about the definition of property rights, a fundamental tenet of liberal democracy, lies at the heart of many land-use disputes. Douglas Baker suggests that disputes over the appropriate allocation of resources will be difficult to resolve when the normative bases for understanding the resource, and the property rights to that resource, are inherently different. Baker uses the example of the Temagami forest in Northern Ontario, an area of significant resource conflict: 'A professional forester described the Temagami White and Red Pine forests with terms such as "efficiency," "optimal cut," "decadent stand," and "allocation units." The same stand of timber was also described by an environmentalist as "remaining heritage," "ecological museum," and "a reserve for our children." '[27] Participants in the conflict, then, are communicating in different languages – that is, if any communication is going on. Can the public interest be determined when participants are not even sharing a similar language when discussing resource values?

This is also the case with terms such as 'property rights' that are bound by both time and space. Concepts of rights change over time and are subject to the interpretation of various communities of interest. Baker outlines various interpretations of private property rights. They include:

- first occupancy and natural rights – 'it's mine because I found it first'
- labour theory – property rights derived from labour invested by an individual (built on a philosophical justification first posed by John Locke)
- utilitarian theory – based on maximizing the interests and overall happiness of society
- liberty – freedom of the individual through property rights.[28]

In contrast to these liberal-classical views of private property, resources can also be viewed as something held in common (as is often loosely applied to air, outer space, fish, and oceans) or a communally-held resource. For example, as Marchak points out: 'Prior to European settlement, B.C. Indians (like many other nonindustrial societies) allowed the environment to be owned and managed by kinship or territorially located groups.'[29]

Baker describes the comparatively recent environmental ethics arguments related to environmental property rights. These range from an anthropomorphic view of the environment (its utility to people) to an eco-centred perspective where the environment has an intrinsic value of its own apart from humans.[30] For some, the liberal-democratic system that supported economic growth has created conditions of ecological scarcity of such a magnitude that a whole new form of governance will be required – one that is much less liberal. Ophuls and Boyan claim that: 'Scarcity, in general, erodes the material basis for the relatively benign individualistic and democratic politics characteristic of the modern industrial era. Ecological scarcity in particular seems to engender overwhelming pressures toward political systems that are frankly authoritarian by current standards, for there seems to be no other way to check competitive over-exploitation of resources ...'[31] Whether or not one shares that interpretation, it is now generally recognized that private property values based on liberal-democratic traditions must share the stage with competing interpretations of the value and ownership of resources. The Canadian political culture itself fosters a philosophy of the collective interest, common property, and representative decision-making. Environmental politics, too, are based on a philosophy of collective rights. In market systems, these rights have usually been subordinated to assertions of private and individualist rights.[32] Yet as Susan Phillips notes, 'The importance attached to the politics of inclusion and the sense of procedural rights exercised by social movements is likely to expand the concept of fairness and place an even greater onus on the state to ensure representation of a full spectrum of interests.'[33] Perhaps the public interest means no more than finding a broadly accepted compromise. If this is the case, the success of consensus-based initiatives will depend on the ability of current political institutions and cultures to accommodate new frameworks for decision-making that are considerably different than those that currently shape our political environment. As Christopher Dunn observes: 'Canadians have come of age politically in the political and constitutional upheaval of the last few decades. No longer do acquiescence and deference characterize the "peaceable kingdom." Canada has undergone a sea change in its political culture and is now at ease with the concepts and instruments of plebiscitarian democracy.'[34]

Democratic Considerations of Public Consultation

Many theorists have examined the role consensus-based processes have played in contributing to democratic goals. If such processes as the Whitehorse Mining Initiative are to work, have a lasting impact, and not end in frustration, the roundtable negotiations themselves must be based on principles that all the participants can accept. One of the principles includes reasoned discourse or discussion among the participants. Such principles are as applicable to broader democratic practices involving the citizenry as a whole as they are to more limited exercises designed to achieve consensus.

New policy approaches such as the WMI have the potential to address the mounting demands of diverse, and often conflicting, interests. These processes may improve information gathering and analysis, thereby generating more effective policy alternatives. These initiatives may be better suited than traditional processes to tackle the persistent, intertwined, and complex policy problems that characterize the policy world in the 1990s. The challenge is to determine what communicative or discursive designs can be modelled to facilitate the interaction of individuals who bring vastly different objectives, attitudes and philosophies, skills, knowledge, and other resources to the table. Multi-stakeholder forums represent one of the most significant innovations in the Canadian political process in recent years.

One definition of 'consensus' is unanimity; in practice, consensus-making involves the balancing of interests and values. Roundtables and other similar mechanisms represent new styles of decision-making and new approaches to governance. There may be clearly defined objectives and the expectation of tangible results. Other important, although less easily identifiable, products of such efforts may also be evident. For example, a series of multiple stakeholder meetings held over a period of time to discuss issues may result in a sense of shared trust and working relationships. These relationships may be called upon in future collaborative bilateral or multilateral relations. There is, therefore, a need for caution in making hasty assessments of multi-stakeholder initiatives.

In resource management, there is a clear need for innovative decision-making approaches and practices that may serve as a bridge for the numerous intersecting, often contradictory, issues that can 'bog down' both public policy and industry officials' decision-making. Roundtable initiatives may serve as a useful forum to help mediate and mitigate disputes between specific conflicting interests in society. Moreover, new consensus-building policy processes such as the WMI could serve a democratic function in a world that is becoming increasingly technologically and socioeconomically complex:

> If we have lost touch with each other as interest groups wrestle for their piece of the action; if we have let natural tendencies for immediate economic gratification discount costs to future generations; if we have lost confidence in our institutions, our leadership, and that leadership's accountability to the public; and if we exclude ourselves from the decision process because the issues are seemingly too technically complex, we may reinforce a self-fulfilling prophecy of democracy's doom ...[35]

The late twentieth-century high-tech, high-speed society is often characterized as one in which citizens are alienated from the decisions that deeply affect their lives. Public consultation is often seen as a necessary way to increase the legitimacy of governments and to appease conflicting public demands. This is particularly the case with sustainable development, where somehow a balance must be achieved between the natural environment and the economy. Roundtables, therefore, constitute a fundamental component of the recommendations of the well-known United Nations Brundtland Commission's report on sustainable development entitled *Our Common Future*. It called for

> a political system that secures effective citizen participation in decision-making, an economic system that is able to generate surpluses and technical knowledge on a self-reliant and sustained basis, a social system that provides for solutions for the tensions arising from disharmonious development, a production system that respects the obligation to preserve the ecological base for development, a technical system that can search continually for new solutions, an international system that fosters sustainable patterns of trade and finance, and an administrative system that is flexible and has the capacity for self-correction.[36]

The Whitehorse Mining Initiative endorses this approach to decision-making. It is the first national mining initiative of this type in the world. It is certainly more ambitious in scope than many other similar consensus-based efforts in other resource sectors. Given its revolutionary approach to mineral policy and its ambitious national design, the WMI serves as an important case study of the viability of multi-stakeholder initiatives. These initiatives are based on principles of public participation; understanding and respect between stakeholders; access to adequate information; integrated decision-making; and new institutional forms based on networks and partnerships.

Conclusion: New Public Policy Approaches and the WMI

The demands on the modern state are leading to the adoption of new policy processes. The attentive public is no longer willing to have the public interest defined through bipartite agreements between governments and narrowly defined elites. Citizens want to have a greater say in decisions made by governments.

Furthermore, there has been a distinctive move toward 'vision statements,' 'mission statements,' and 'statements of commitment.' Such statements can be found in corporate and government annual reports and in natural resource roundtables: 'These brief statements are designed to instil a central purpose in all employees as they go about their work. Government departments and agencies soon jumped on the bandwagon holding staff meeting after staff meeting to define a central mission to which all employees could relate. Once again, the client was invariably put front and centre.'[37] Service to the public and a responsive, open bureaucracy are all goals that fit nicely into the newly emerging policy world. Yet once again, the unresolved issue remains: who are the clients and how are their interests to be reflected in the policy process?

An examination of the Whitehorse Mining Initiative will highlight some of the policy issues involved in attempting to introduce a new policy approach to the management of natural resources. The initiative will be considered in the context of whether it is perceived to be a 'legitimate' approach to decision-making; whether the Canadian public interest can be advanced in such a process; and whether the process does achieve some measure of success in terms of developing a more harmonious and sustainable approach to mineral development in Canada.

For multi-stakeholder consultative processes to be considered legitimate, several issues must be addressed. First, such processes will have to be devised in a way that is particularly sensitive to representation. Participation will have to be inclusive enough so that the ultimate policy will not be viewed as one that has been 'captured' by particular interests. Ironically, if the range of choice is broad enough to be representative of the diverse interests in society, it will be difficult to achieve consensus. By virtue of the selection process itself, the participants will not even share the same system of values. All they may have in common is a recognition of the need for compromise. Moreover, each of the stakeholders needs to possess the requisite human resources, technical expertise, and skills to work in consensus-building partnerships. Without adequate information, knowledge, and skills there is a possibility that the result may be unsatisfactory – not necessarily the best solution, just one about which all participants could agree. Moreover, resource constraints or lack of experience may limit the scope for innovative thinking and strategic alternatives.

Second, the reasons for assembling a particular group of participants must be clearly stated. Will the decision-makers know what they want of the participants they have selected? Are individuals gathered together for their expertise on the subject area? Do they serve as a small sample of a more generalized public opinion? Are participants expected to be representative of other interests? Perhaps public participation is thought to be a good thing in general, and the more of it the better! Liora Salter raises a number of perceptive points in her article on the democratic potential of regulation:

> In practice, few regulators have really decided what they want to achieve through public participation. Having not decided what expectations of participation are reasonable and appropriate in their specific case, they fail to set clear guidelines to encourage it ... attempts to encourage direct public involvement in decision-making fail when regulators are themselves confused about what they want to achieve from direct democracy.[38]

Third, the issue of objectivity must be addressed. Can the advisory group maintain its objectivity in order to provide a representative perspective of the various interests? Susan Phillips notes the valid, and often stated, concern that participants in consultative processes might be captured by the agency or the process itself: 'Should a group become involved in the on-going process of co-management, it becomes extremely difficult to step outside the relationship to be a vocal critic without alienating the co-management partners.'[39] She suggests that an even greater danger may be that effective participation will be limited to those who possess some knowledge of the area to be regulated.[40] Students of politics have long been aware that those who participate are not necessarily representative of the broader population.

Fourth, the process itself will have to be open enough so that participants feel that they contributed to the shaping of the policy, not that they are simply endorsing a previously determined plan. Their feelings of efficacy are crucial to the ultimate success of the process.

Fifth, will the sizeable increase in the number of consultation mechanisms consume scarce time and resources of participants in a way that will limit their effectiveness and interest in the processes? In an era of budget restraint and fierce competition for support base, can participants afford the cost of participating in such consultations?

Finally, criteria must be devised to have some way of assessing whether or not a process has been successful. Identifiable goals will have to be set out with deadlines and time limits so that participants will have some idea of what it is that they are working toward. Certain identifiable goals may be met by the participants. In the case of the mineral industry, a more predictable policy environment that contributes to a healthy mining sector is the bottom line. Answers to more fundamental questions, however, such as whether these processes will be democratic, legitimate, and in the general public interest, will undoubtedly prove elusive.

The Whitehorse Mining Initiative was quite successful in meeting some of the above criteria in terms of the immediate goal of reaching an accord. Its long-term future as yet remains unresolved. Many suggestions were made during the process, however, that could help map out future directions in this new relatively uncharted territory in consensus-based consultative policy-making. The 'art of the possible' has not yet received its due, at least in the latter half of the twentieth century. If decision-makers do not identify the problem before them accurately or adequately, and if alternative solutions do not receive their due consideration, then it is not surprising when policy decisions do not always alleviate the problem they are expected to solve. Land-use planning in the 1990s is characterized by a rapid adoption of public consultative roundtables in attempts to reconcile widely diverging perspectives about appropriate resource use. As participants become more familiar with this new policy style, such a process may become publicly recognized as an effective and more legitimate way to proceed with land-use planning. The possibilities are there.

Currently, however, advocates and participants in these consensus-based processes are still undergoing a steep learning curve. This was certainly the case with the innovative Whitehorse Mining Initiative. For many of the participants, the process was unfamiliar and, in some cases, unwelcome. The wariness and scepticism of the participants, and the lack of preparation, plagued the process. Yet many of the participants engaged in intense discussions to try to reach a consensus. This new form of interaction led to something that surprised many of the sceptics – a signed agreement.

2
Assessing the Situation: Challenges to the Mineral Industry[1]

The Whitehorse Mining Initiative was created in response to the Canadian mining industry's concern that its continued viability was threatened. The Canadian mining industry continues to be an important contributor to the Canadian economy and a world-leading producer and exporter. Public values and attitudes toward resource development, however, are changing. Public perceptions are being influenced by shifting demographic patterns, environmental values, the emergence of a variety of competing land uses beyond resource development, and rapidly transforming political, economic, and physical environments. Some political economists and others question whether Canada should continue to promote its staples industries at the expense of diversifying into the secondary and tertiary sectors. The mineral sector faces stiff competition in the global economy, whether it comes in the form of competition from new materials – such as ceramics or composites – or from other mineral competitors in places such as the former Soviet Union, Latin America, or Southeast Asia.

While the Canadian-based mining companies continue to compete in the world marketplace, they face a number of domestic political challenges. With no clear picture of the public interest, governments are buffeted by the growing demands of competing interest groups who do not hesitate to articulate their particular visions of the public good. Rising pressures for public participation coupled with escalating land-use disputes, such as the 'Windy Craggy' controversy in northwestern British Columbia, have led to increasing levels of mistrust between members of the mineral industry and other interest groups. Such events do little to inspire investment confidence in the resource sector. The growing complexity of the modern state limits the ability of governments to provide a coordinated, regulatory environment that would attract investment capital to resource-based projects. Industry representatives, however reluctantly, are now beginning to accept that the policy environment will not be favourable to the mineral sector unless compromise, communications, and negotiations take place with other groups and interests who influence the decision-making environment. Multi-stakeholder approaches, integrated resource management, and the Whitehorse Mining Initiative are all attempts to introduce an element of stability and consistency into resource policy-making.

Developing Consensus: The Whitehorse Mining Initiative

The regulatory and competitive challenges facing the sector stimulated a few farsighted mining representatives to develop the Whitehorse Mining Initiative and to sell the idea to members of their industry, as well as to other diverse members of the resource policy community. It was not, and still is not, an easy sell. Not only does the industry have to persuade policy-makers of the continuing contribution of mining but they must also agree among themselves on a strategy about how this is to be accomplished. To be politically astute, the mining sector collectively needs to understand and accommodate interests of other political players. This is not something that comes easily to those in the industry. Prospecting and developing, in particular, is a solitary and, at times, secretive pursuit. Certainly at the exploration stages, the initial process of discovering and staking one's claim to an exciting prospect is not one that is immediately broadcast to potential competitors. It is worth noting that this is not a cultural trait peculiar to Canada. One observer of the American mineral industry suggests that it, too, has a long way to go:

> I propose that the historic culture of the mining industry, even more than of other manufacturing sectors, has been one of stubborn isolation. That culture grew as a response to very real physical challenges. It was, at its origin, realistic. But it is realistic no longer, and by resisting integration into the norms of society at large – and I mean particularly environmental norms – the mining industry is vainly resisting an inevitable trend.[2]

The Canadian industry is in a similar position. If industry representatives wish to be effective politically, they must figure out a way to persuade members of their associations of the necessity of a cultural and attitudinal shift. Industry must be able to recognize the interests and concerns of competing resource users as well as the complex role governments must play in balancing the interests of the economy and society. Senator Lowell Murray advises:

> Compromise and conciliation have ceased to be fashionable in an age typified by media confrontation and pressure group politics. But their lack of appeal doesn't render them any less critical to resolving the conflicting demands of different regions, peoples, or interests. Real democratic governance occurs when decisionmakers feel able to exercise their responsibilities

> in the broadest possible interest – the public interest. We all share that responsibility – to listen, to reflect, and to seek solutions that satisfy not simply the demands of the day but the needs of the people ... The point I would make to you and to other sectoral or industry groups is that you pursue your legitimate interest in a broader context.[3]

When groups such as the prospectors and developers or the provincial mining associations try to influence the political process, they often fail to differentiate between political strategy and a public relations and communications campaign. As W.D. Coleman points out, political strategy encompasses an ability to know how to develop alliances with other groups, examine who wins and loses by a particular proposal, and attempt to foster 'win-win' situations.[4] The process of influencing public policy and introducing new legislation is no longer a simple matter of convincing a minister of mines to take and sell one's ideas to the cabinet or premier. The policy approval process is complex. Proposals must be vetted through a number of committees that must evaluate the proposal; they must be sold to competing departments, coordinated with the activities of other orders of government, and, frequently, examined by outside interest groups.[5] Coleman also notes that the industry is a collection of loose associations that include prospectors and developers, associations devoted to particular commodities, and provincial and national mining associations.[6] These groups have yet to learn how to work in a concerted fashion. Before they can persuade others of their concerns about the mining industry, they need to be able to come to terms internally. If these associations can learn to work more closely together, they might be able to develop some self-discipline that will allow them to play a more effective role in the public policy arena. As noted above, many prospectors and developers, in particular, were attracted to the lifestyle for the independence and freedom that it offered. These individuals may be the least prepared of any to work in the new proposed environment.

On the other hand, if the mineral sector is able to build on the innovative WMI accord, and move toward a long-term coordinated strategy, it may find itself a world leader in a way that goes far beyond the production of minerals and technical expertise (as historically has been the case). The Canadian mineral industry is in a position to demonstrate international leadership by showing that it is possible to forge a consensus between diverse interests in order to foster a cooperative climate for responsible and sustainable mineral development.

As the following discussion illustrates, however, despite its contributions to the Canadian economy and society, the industry is facing a number of serious challenges that it must address if it is to continue to play an important role in Canada and in the international arena.

Competitiveness and the Canadian Mining Industry

Mining and mineral processing have played an integral role in the development of Canada's economy and society. The mineral industry has been referred to as the 'backbone' of many regional economies in Canada. Furthermore, many communities rely on mining-related activities in the secondary and tertiary industries (including tourism, transportation, manufacturing, and services). The mineral sector includes companies involved with mineral exploration, extraction of ore, milling/concentrating, smelting, refining, and processing. Within those broad categories, it has a sizeable mining-related service sector that includes remote sensing and geodata processing, aviation services, mining instrumentation, etc. Canadian mining is increasingly high tech and is developing an international reputation for its technical expertise and research in automated and robotic mining, and environmental technologies that are used to mitigate the impact of acid-mine drainage.[7]

The mineral industry remains an important pillar in the economy. Canadian (nonfuel) mineral production for 1995 totalled $19.3 billion, an increase of 15.5 per cent over 1994. Overall, the minerals and metals industry constituted about 4.4 per cent of Canada's GDP in 1995 ($23.7 billion). The mineral industry employed over 300,000 individuals in jobs ranging from mining to metal fabricating.[8] Canada produces over sixty mineral products. It ranks as one of the top five countries in the production of seventeen different minerals. It is first in the production of potash, uranium, and zinc; second in nickel, sulphur (elemental), asbestos, and cadmium; third in platinum group metals, copper, titanium concentrates, gypsum, and aluminum; fourth in molybdenum and cobalt; and fifth in lead and gold.[9]

The industry experienced a significant drop in world metal prices and a downturn in exploration in the early 1990s (see Figures 2.1 and 2.2). Between 1986 and 1991, Canada was unable to attract any very large new mining projects (i.e., those that had a capital cost over $250 million) while Latin America attracted five. According to a financial analyst for Cominco Ltd., mining investors are now more attracted to South America and Southeast Asia. This is not due to lax environmental regulations, it is argued, but because of 'the streamlined approval processes and enhanced investment climate.'[10]

By the middle of the 1990s, the mineral outlook began to improve. In 1994, for the first time since 1989, more mines opened (twelve) than closed (nine). Metal prices also improved. Exploration expenditures in Canada were also up, with Canada commanding 13.6 per cent of the worldwide exploration budget in 1994, in third place after Australia and the United States. The 1994 figures of $625 million (up from $477 million of the previous year) were expected to continue to rise in subsequent years.

Figure 2.1

Exploration expenditures in Canada by junior and senior companies, 1969-94

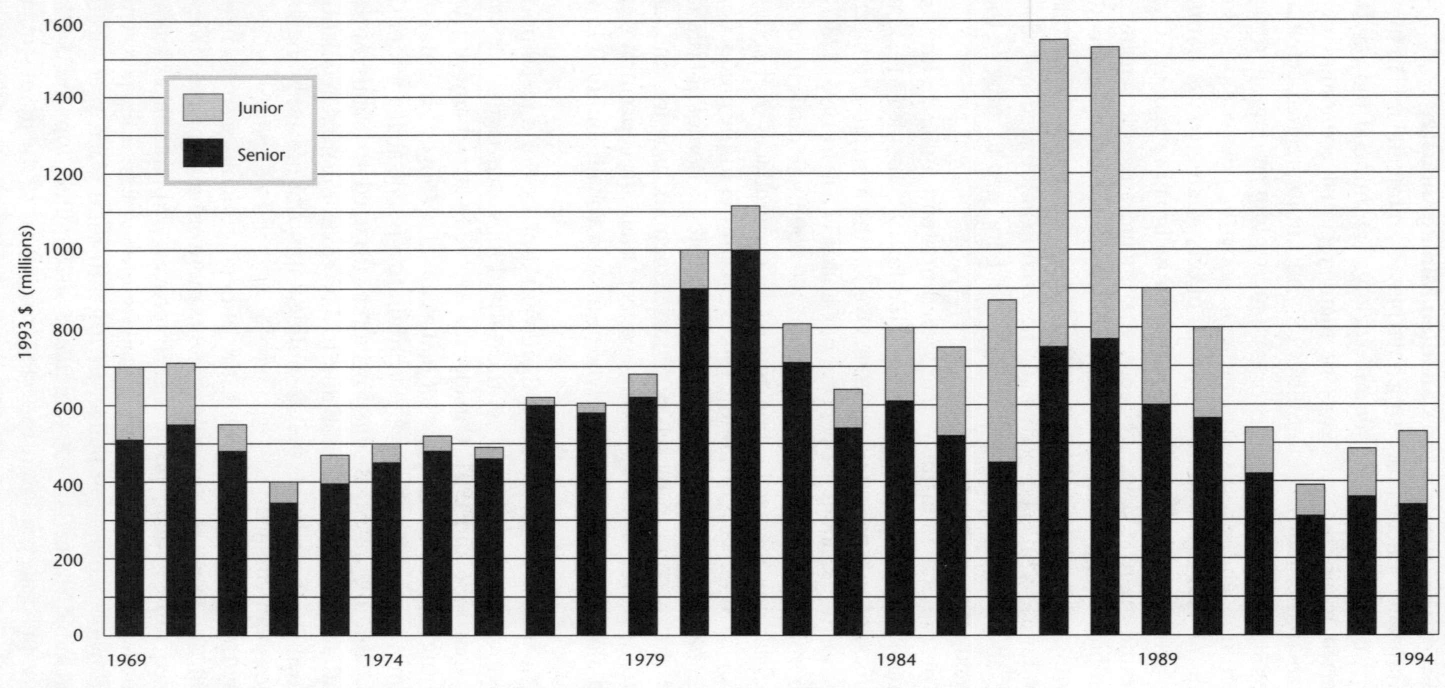

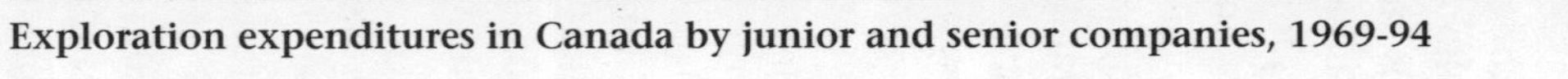

Sources: Mining Sector, Natural Resources Canada and Federal-Provincial Survey of Mining and Exploration Companies.

Figure 2.2

Worldwide exploration budgets, 1994

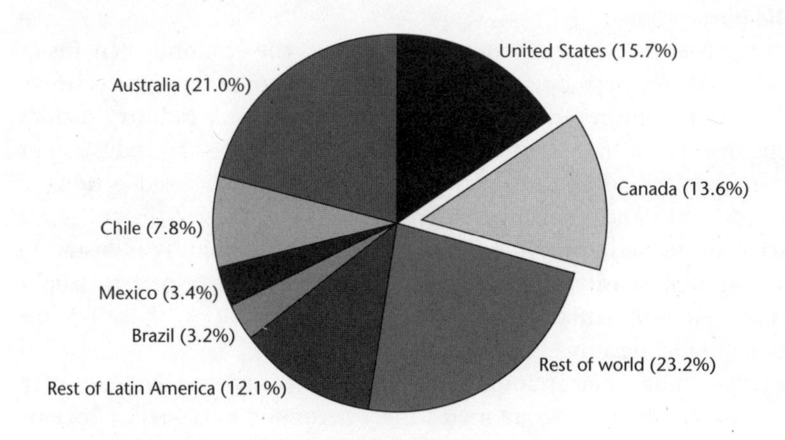

Sources: Metals Economics Group; Natural Resources Canada.

Through innovative practices and automation, the mineral industry has made significant improvements in productivity and increased its production levels. George Miller, president of the Mining Association of Canada, stated that 'although its [mining's] share of GDP has dropped, we've held our own in quantity of production while the Canadian economy has diversified around us.'[11] Although they are world leaders in mining, members of the Canadian mineral sector must nevertheless contend with a diminishing share of the global market and decreasing known reserves of minerals. Mineral economists have pointed out that the known, easily accessible economic deposits are becoming depleted in Canada. Many large mines could be closed in the next decade without new mines to replace them. One executive of a prominent mining company echoes many others when he states that there are still excellent opportunities in Canada if the regulatory environment were hospitable. Recent significant discoveries in the Northwest Territories and Labrador lend support to his assertion. While Canada still attracts 13.6 per cent of the worldwide exploration budget, Chile now captures 7.8 per cent of overall expenditures (see Figure 2.2). Furthermore, governments in developing countries often pursue aggressive marketing strategies to attract foreign investment. In addition, rich ore grades and excellent geological opportunities in developing countries provide a very attractive environment for investment.[12] These 'pull' factors are clearly important and should not be underestimated. On the 'push' side, in Western liberal democracies in general, political institutions and the public policy environment have been changing, creating a less hospitable environment for resource-based activities. Canada is no exception.

Public Perceptions

Much has been said in the popular media about the economic benefits of the so-called sunrise industries that compete for space on the electronic highway to the future. In contrast, a traditional resource industry such as mining does not attract as much public attention – unless an industrial or environmental accident takes place. At such times, public recollections of poor historical mining practices often rise to the surface of the collective conscience. Furthermore, when some members of the mining industry do behave in an insensitive or irresponsible manner with respect to labour concerns, environmental practices, or local communities, those actions continue to fuel negative public perceptions.

Negative public perceptions of development may also be part of a broader phenomenon. We are increasingly becoming a risk-averse society, particularly in the case of the unknown. While individual citizens are willing to take significant risks with something familiar, such as driving on highways, they are less willing to accept something unfamiliar where the risks are difficult to calculate, as might be the case with a new mining project. The public is also much less likely to trust scientific reassurances when they know that scientists often can, and do, disagree in their conclusions.

Members of the urban public are also often ready to discount the relevance of mining to Canadian society. An unaware public and an unsympathetic policy regime have an adverse impact on traditional industries. It should be noted that this perception extends beyond mining. There is limited public recognition of the continuing importance of resource industries to Canada in general. One observer of the Canadian economy noted that: 'Perhaps the most important factor in Canada's ability to move forward is the attitudes and the mindset of individual Canadians ... Canadians must better understand the foundations of their past prosperity, and the fact that the comfortable order is disintegrating.'[13] Political economists have often argued that the decline of Canada's staple-based economy was inevitable because of its failure to diversify, relying instead on the abundance of its natural resources to sustain its wealth. Further, they question the economic and social value of developing policies that support primary industries as opposed to secondary manufacturing or the service sector.

Primary industries do continue to contribute considerably to manufacturing and service activities in Canada. In addition to having access to lower-cost raw resources, secondary industries have been developed and shaped in response to our resource-dependent economy. This includes not only the obvious examples, such as transportation facilities or business-related tourism, but also new industries related to information technology.

For example, the Canadian satellite industry, which has achieved that peculiarly Canadian label, 'world-class,' was developed to support many resource-related activities, including mobile communications, remote sensing for resource management and surveillance, weather forecasting, search and rescue, and navigation.[14]

There are, however, many who do not support the view that secondary manufacturing necessarily depends on domestic primary inputs. Rather, it has been claimed that such assertions were more persuasive in the early 1970s when there was a belief that raw material shortages reflected permanent conditions of global scarcity and firms quickly moved to secure supply. Robert Young and others argue that this situation has now changed: 'In recessionary times, it is not access to resources that counts, but access to markets, and this renders primary industries dependent on the fortunes of their downstream customers rather than vice versa.'[15] One has to be careful about accepting such a generalized statement at face value because access to resources is still important to secondary industry, particularly those minerals used for building materials or fuel. There is, though, no question that the mineral industry faces many competitive threats. The argument that primary industries are dependent on their customers remains valid.

New Competitive Realities

As the composition and demands of markets change, the mineral industry must also be prepared to apply a diversity of strategies to retain its market share. In addition to competition from low-cost producers, another significant threat to the mineral industry has been the introduction of new materials, such as metals, ceramics, and engineering polymers.

In 1991, a consultation paper was released by the federal government entitled 'Prosperity through Competitiveness.' The paper was part of an overall initiative to determine ways to facilitate Canada's continuing prosperity. The discussion paper stressed the importance of increased investment in science and technology: 'Canadian industries, including resource sectors, invest proportionately less in R&D than their counterparts in other leading industrial nations. Meanwhile, our achievements in fields such as telecommunications, nuclear technology, and medical science show that we can lead the world when we focus our efforts.'[16] The report recommended that competitive strategies should build on Canada's traditional areas of economic strength – namely, the resource sector. The report cited the case of INCO's extensive initiatives in R&D, ranging from new and more efficient processes for ore extraction to automated machinery that has marketing potential.[17] Canadian mining engineers are working assiduously to develop a completely automated mine, one that operates without underground miners. INCO has developed a multi-vehicle load-haul dump

(LHD) system that is operated by remote control (teleoperation) through the mine communication system and video link. Other initiatives include automated rock monitoring with laser, sonar, geophysical, and image-processing technology; and automated explosives handling, which includes the use of robotic arms to load blastholes.[18] The development of such new technologies, which build on Canada's traditional strengths in the resource sector, might be viewed as the only practical route to diversification and growth of the country's economy.

There are many approaches that can be used to improve the competitiveness of the industry. They include increasing the level of R&D investment, applying high-technology solutions to reduce costs and/or reduce environmental hazards, educating and attracting highly skilled workers, improving marketing techniques, developing investor incentive packages, etc., and developing a broader political awareness. Global and national politics have become increasingly interdependent and complex. The traditional focus on exploration and development through to mining, milling, smelting, refining, and marketing of minerals will no longer be sufficient to keep the industry competitive. If the mineral industry is to be perceived as up-to-date and competitive with so-called 'sunrise' industries and be successful in the international competitive arena, it will have to adopt a different strategy. The strategy will have to be one in which its leaders are sophisticated not only in terms of mineral development and new technologies but also in terms of broadening their awareness of the changing global and domestic economic, social, and political arenas.

Mineral Scarcity?

In addition to a diminishing share of the global market, the Canadian mining sector must contend with decreasing known reserves of economically viable minerals. Such trends lead policy-makers and others to revisit classical economic assumptions based on doctrines of 'increasing natural resource scarcity and diminishing returns, and their relevance in the modern world.'[19] What is meant by 'scarcity'? Certainly, as consumer demand multiplies, so too does the demand for natural resources. As a result, there has been a dramatic global impact on the physical environment. Depletion of the fisheries, air pollution in metropolitan and industrial areas, contamination of ground and surface water supplies, and the rapid reduction of the world's forest cover all point to a growing scarcity of vital resources. The blanket use of the term 'scarcity,' however, can be misleading – particularly if it is applied to the available supply of minerals themselves. Few outside the mineral industry are aware of the difference between 'mineral reserves' and 'natural endowment.' The failure to distinguish between the two can lead to the perception that 'depleted reserves' are synonymous with 'depleted mineral resources.' Mineral economist Michael Doggett explains:

> Mineral resources are placed in one of three overlapping categories. The first is 'reserves' which are known, economic at present costs, and feasible to mine with current technology. The second is known and unknown 'resources' which are potentially economic at relevant future price and cost levels, and feasible to mine under current and future technology. The third category is the 'resource base' containing both known and unknown material which may be economic or uneconomic under current or future conditions of prices, costs, or technology.[20]

A change in world prices, an improvement in extraction technology, or a supportive regulatory environment could all lead to an increase in mineral reserves.

There is a general lack of awareness about the increasing productivity in the mineral industry as well as little understanding about the potential of minerals themselves. One analyst has argued that economic mineral reserves could be viewed as a flow rather than a 'limited stock.' Margot Wojciechowski claims that:

> The flow is maintained by investment: investment in exploration; and investment in new technology ... innovation means new discoveries; it means major expansions of reserves and mine life at existing mines; it means that deposits that formerly were useless waste rock become economic; it even means that old waste dumps and tailing piles become new mines ... Metals mined become part of a permanently recycling and constantly growing stock.[21]

Prospectors and developers argue that those making the 'scarcity' arguments focus on what they know of the discovered parts of the mineral world; they ignore, deliberately or otherwise, the potential of mining's vast 'terra incognita.' Also ignored is the fact that, unlike organic commodities, metals are not consumed when they are mined and used. Scrap metal can be recycled and used again.

Scarcity theorists point to well-founded concerns about the depletion of known reserves, lower ore grades, and dying single-industry towns, issues that are all highlighted by the media. Yet given the virtually untapped resources of many parts of Canada, particularly the North, the industry claims that the surface of mineral endowment has just been scratched. This argument has been most recently supported by the huge deposit discovered in Voisey's Bay, Labrador. Aptly named 'Discovery Hill,' the deposit is described as one of the 'most dynamic and back-to-basics mineral discoveries' made in Canada in a long time. The area is estimated to contain $10 billion of high grades of nickel/copper and cobalt in a near-surface, massive sulphide zone.[22]

In spite of promising discoveries, opening up new regions is fraught with difficulties, some political, others economic and technical. One difficulty lies in discovering, developing, and mining *economically viable* deposits. Mineral producers and investors will leave to go offshore if they believe that the costs of mining a deposit, even a high-grade deposit, prove to be too high in Canada.

Mineral supply consists of several stages, from primary exploration to production of mineral commodities. Mineral economists Brian Mackenzie and Michael Doggett point out that in the mineral exploration stage, 'all finding, development, and production activities represent controllable choices. This is the only time a mining company is not committed to particular programs, projects, or mines.'[23] It is at this stage, the authors suggest, when there are no sunk costs, that corporate exploration decisions are particularly sensitive to uncertainty and changes in government policy. A large part of the political uncertainty can be attributed to changing taxation policies, changing environmental and health and safety regulations, the length and uncertainties of the environmental review processes at both the federal and provincial levels, and land-use policies.

Currently, for example, the industry expresses concerns about policies surrounding 'protected' park lands and wilderness areas as well as the unresolved state of Aboriginal land claims. They suggest that the 'legal limbo' surrounding the land affected by the claims process means that even where there is no legislation prohibiting exploration, neither are there any guarantees to access or to mineral rights.[24] Members of the industry maintain that as long as the land claims remain unresolved it will be difficult to attract capital to these regions for high-risk discovery and development projects. It is argued that land-access controversies, including uncertainty over the meaning of mineral title, may contribute to the pressures that drive mining investment out of Canada to countries where the investment climate is more attractive.

It is difficult for the interested observer to determine the extent to which investment is leaving the country as a result of regulatory uncertainties in Canada's land-use policies and how much is due to other factors. Many analysts will point out that these concerns are by no means confined to Canada. Other Western resource-dependent countries, such as Australia, must wrestle with similar issues.[25] In choosing where to explore for minerals, the investor must consider the overall investment climate, which includes the availability of a skilled workforce, the accessibility of high-grade ore deposits, labour and production costs, and regulatory considerations. One mining executive identified several of the reasons that mining companies decide to move to so-called 'frontier areas' and selected areas outside Canada:

> Companies want to focus on areas with the best potential to be endowed with large, high quality mineral deposits; and the largest and richest

> deposits will be the first detected and developed. There is a better chance that these will be found in relatively unexplored areas ... On the push side, we see in Canada substantially increased risk to doing business and diminished return through increasing levels of taxation, increasing regulation, and decreasing land availability ... Management of the issues of regulation, taxation, and land availability must begin with the very first steps of exploration either in Canada or any other political jurisdiction.[26]

In addition to those factors, ideology may also be playing a role in influencing the perception of investors. The culture of the mineral industry is not one that readily embraces the idea of the activist state. When encountering a new regulatory regime that the industry considers unnecessarily restrictive, members of the mining sector will push back. The notion that Canada's regulatory policy environment is driving investment away is widespread. How much weight to give this factor, however, is a subject for debate. Is industry's perception that Canadian governments are creating an unfavourable environment for mineral development contributing to the more cautious investment climate? If this is the case, the Canadian mineral sector's ability to remain competitive will depend on how well it will be able to forge alliances with governments and others who influence policy-making. Only by doing so will the industry and others be able to distinguish perception from reality, and from that point develop cooperative strategies for sustainable mineral development.

In sum, scarcity theories predicated on assumptions about the depletion of Canadian ore reserves do not take into account the 'renewability' of the resource base and the potential offered by Canada's mineral endowment. Nevertheless, the Canadian industry is still concerned about its depleting stock of mineral reserves. The finding and development of new mineral deposits in Canada are not only influenced by geophysical, economic, environmental, or technical variables but also by the political and regulatory environment. Furthermore, the federal, pluralist nature of the Canadian political system has led to a fragmented policy-making environment. Such an environment may help ensure that resource-use decisions are carefully considered from a variety of perspectives, leading to a more integrated resource-management approach. From the industry perspective, however, a lack of certainty in the investment climate (e.g., fear of expropriation of mining claims or changing regulatory requirements) will be a factor in determining whether or not Canada can attract capital for mineral exploration and development.

Politics and the 'Embedded State'

The Canadian mineral industry has often looked to technological advances rather than to the political and economic environment to find answers to its competitive difficulties. The largest proportion of research money is devoted to discovering more cost-efficient ways to extract ore and produce minerals. The industry has not paid adequate attention to the public policy environment, which is seriously beginning to affect its ability to remain competitive. It has ignored the social and political forces crucial to its continued survival. Moreover, there is an industry-wide awareness of 'government-mandated costs,' but that is not accompanied by an understanding of the role governments must play in regulating the economy and in providing a stable and healthy political and social environment essential to a prospering economy.

In recent decades, state-society relations have become increasingly intertwined. Alan Cairns refers to this phenomenon as the 'embedded state' and suggests that the Canadian federal state 'has become a sprawling diffuse assemblage of uncoordinated power and policies, while the society with which it interacts is increasingly plural, fragmented, and multiple in its allegiances and identities.'[27] As the 'embedded state' develops more fully, many groups are increasingly able to exercise an effective 'veto' on the policy process. The federal political structure facilitates a policy environment that includes both competing producers and consumers, and competing governments. Defacto 'vetoes' on government policy are achieved through the ability of various groups to recruit corporate and government opponents. The ability of groups to capitalize on the fragmented nature of political authority does not contribute to a stable regulatory regime. Given this policy environment, the fact that the public and many policy-makers seem indifferent to the need for continuing competitiveness of the mineral sector is viewed as a fundamental problem for the industry.

Access to land for exploration and development is becoming a very contentious issue for the mineral industry. Issues dealing with questions of land access serve to illustrate the growing conflict between the interests of industry and other groups. The industry needs access to large tracts of land in order to explore for new economically viable deposits. Other groups are now making competing claims on many of those areas. Cairns's embedded state is 'tied down by multiple linkages with society,' limiting its manoeuvrability.[28] In this environment, the mining sector needs public support and awareness of its basic requirements.

Interest group studies, public policy, and institutional analyses all point to the influence of the public in shaping government decision-making. As Cairns and Williams state: 'The ever more encompassing coverage of the planet by the state system, now extending to the oceans and the stratosphere, has been accompanied by a tightening of the links between citizens and individual governments. In the Western world increasing global interdependence has coincided with a general enhancement of citizenship as a private status invested with rights and duties, relative to more private definitions of self.'[29] Cairns and Williams suggest that in contemporary society, a plurality of identities has emerged, with groups enjoying 'considerable success in expanding discussions of human rights and in securing positive government responses through a wide range of public policies as well as through the legal and constitutional validation of various rights claims.'[30] Governments are now recognizing the needs, interests, and some of the historic claims of Native peoples, as well as public demands for a healthy, safe environment or for the right to participate in decisions that directly affect them.

Historically, (as noted in Chapter 1) the resource sector did not need to share the political agenda-setting arena with other groups. The public policy environment was very supportive of the mineral industry. Policies were geared toward the development of resources, an essential part of Canada's staples-based economy. Furthermore, as H.V. Nelles explains, the early stages of mining regulation were primarily based on 'a question of obtaining an adequate public return from the development of what was thought to be a public property.'[31] Marsha A. Chandler describes the early relationship between state and private interests: 'The traditional arena that gave rise to these policies can best be characterized as narrow and clientele-based, a symbiotic relationship between the provincial government and private-sector interests. Organized along sectoral lines, government departments that dealt with mining interests did not have to compete among themselves; nor were there any significant challenges from outside.'[32] This relationship is now being challenged both internally and externally. The economy has diversified, and in spite of productivity gains, mining's economic and, therefore, political impact has decreased; it does not contribute the same share to provincial revenues that it once did, nor does it generate as many jobs.[33] As a result, its relative importance on provincial agendas has diminished. In addition, in the early years of economic development, there was relatively little competition for the land that was leased to the primary industries. However, other interests, particularly environmental and Native groups that value Canada's natural resources for something other than their economic worth, are pressuring both the federal and provincial policy-makers to recognize those values.

Most stakeholders would like to see a consistent, stable planning approach to the management of our natural resources. This goal, however, is difficult to achieve because of the pluralist, federal structure of the Canadian political system that lets both environmental and business groups play one government off against the other. According to Robert Presthus, 'This peculiar version of the 'separation of powers' enables interest groups to penetrate and counter the national system very effectively by providing alternative provincial bases of power and avenues of access for such groups.'[34]

Ottawa has constitutional authority for Native peoples and lands, fisheries, and navigable waters. In January 1992, the Supreme Court of Canada dismissed an appeal launched by Alberta opposing federal 'intrusion' into provincial jurisdiction over resource management and land use. The court ruled that the federal government must conduct an environmental review of Alberta's Oldman River Dam project and others like it. As a result of the Oldman Dam decision, all large-scale projects that have an impact on these areas must now be subject to federal environmental assessments. Big projects typically must pass assessments conducted by both levels of government. Unless provisions for a joint federal-provincial environmental assessment process are effectively implemented, the time-frame for project approvals could be lengthened considerably.[35] As George Hood has noted, 'The rules and procedures in environmental assessment are sufficiently vague that any degree of opposition may be enough to prompt a more detailed and costly review than would be legitimately warranted.'[36] For those concerned with the potential negative environmental impact of a large development, the ability to pursue a variety of avenues of appeal can be an asset if one route proves unsuccessful. If, however, the goal is to achieve issue resolution, intergovernmental conflicts can lead to confrontation politics and result in deadlock.[37] The uncertainty caused by federal-provincial disputes over the environmental policy arena, and the rapidly changing regulatory environment (sometimes referred to as the 'moving goalpost scenario') are adding significantly to industry's costs and its concerns.

Structural forces within the Canadian political system have served to exacerbate tensions and to impede efforts to achieve a compromise between competing interests. Forces such as overlapping federal-provincial responsibilities, the entrenchment of the Charter of Rights and Freedoms, and incremental policy-making all serve to obstruct the drafting of regulations that would satisfy either environmental or business interest groups. The segmented political system provides various groups with multiple channels to influence the policy process – it also serves as an impediment to groups in their ability to rally and lobby effectively. Either way, the system often discourages development of an effective, coordinated response to many competing concerns.

Land-Use Policy and Mining: The Case in British Columbia

Land-access issues, referred to in the previous section, highlight the difficulties inherent in the Canadian federal system, as governments attempt to formulate acceptable policy approaches to resource development. Although all provinces are confronted with similar issues, some of the most contentious areas lie in British Columbia. In BC's resource-dependent economy, the $2.64 billion per year solid mineral industry (excluding petroleum and natural gas) produces commodities that constitute over one-fifth of British Columbia's total exports and generates considerable spin-off economic activity. On the other hand, many people concerned about environmental issues related to mining oppose policies that would permit mineral exploration and development in sensitive ecological, wilderness, or park areas.

Parks Plan '90 was announced in June 1990 for the purpose of drafting a provincial land-use plan. The plan identified over 100 areas as potential candidates for park designation. During the ensuing public consultation, meetings were held in an attempt to accommodate many competing views in the province, including those of mineral developers who were alarmed by the size of the areas recommended for withdrawal from resource development.

The 1991 BC provincial election saw the New Democratic Party (NDP) elected to form a new government. During the election campaign, the party stated its support for integrated, multiple use of British Columbia's land base with the exception of specifically designated parks. It also supported the 1987 recommendation of the World Commission on Environment and Development (Bruntland Commission) to expand protected areas to 12 per cent of the total land mass (a position also shared by the federal government's former Green Plan). The party said that it would not allow mining or exploration on land designated for park or wilderness use.[38]

Industry representatives had difficulty with this policy, suggesting that protection means to defend against danger or injury, not isolation or removal.[39] They asserted that exploration itself has a relatively benign effect on the environment and that any affected land is easily restored to its original condition. Moreover, it has been estimated that only one in 5,000 prospects reaches the production stage. Further, mining itself occupies less than one-tenth of 1 per cent of BC's total land base. This argument, however, does not include the 'shadow' effect of mining, which includes the roads and the secondary development that takes place alongside the mine. In response, industry will point out that when compared with agriculture and forestry, mining causes only a fraction of the land disturbance.[40] From this perspective, mining interests object to having the same land-use restrictions placed on them as is the case with forestry. With land-use decisions that are often determined with the forestry sector in mind, the unique requirements and effects of mining may tend to get overlooked.

Industry representatives state that it is difficult to predict what parcels of land will have the most potential for economic mineral development and therefore view the establishment of protected areas as a particular threat to mineral exploration:

> The miner ... really doesn't have a clue where he will find an orebody. He knows which are prospective areas at any given point in the development

of technology and he has a fuzzy but continuously improving understanding of the geological forces that create ore, but he doesn't know where his next mine will be. Having found it, he cannot move it, however much he might like to. Trading areas just gives him mossy pasture in place of ore.[41]

Many northern or alpine regions may be potentially high in mineral resource value, but prospecting remains an inexact science. An array of technologies help determine the location of the most economically promising deposits. Tools include remote-sensing satellite technology and computerized mapping generated from geophysical and geochemical data. Sophisticated geographic information systems are valuable; they do not, however, replace exploration work, nor can they adequately forecast the market environment.[42] Scientific developments in theories of ore formation, improvements in recovery technologies, or changes to the regulatory or market environment could render a formerly worthless deposit valuable.

Calls for designating land solely to be used as wilderness preserves pose a particular concern to the mineral industry because of its unique requirements to explore large areas of land in order to discover very few new mineral deposits. A great deal of money goes into the early high-risk exploration stage. Mackenzie and Doggett have estimated that the average exploration costs for an economic base-metal discovery between 1946 and 1988 amounted to $64 million, with the average exploration period estimated at eleven years during the same period of time.[43] One mineral industry representative suggests that the public and governments are not attuned to this difficulty:

> A company in the manufacturing sector can build a new plant where it wants to, and invests relatively little before construction begins; that is, substantial investment does not take place until after the environmental assessment process. For someone with this model in mind, it is difficult to understand that a mining company cannot initiate the environmental assessment process until after the expenditure of a substantial sum for finding and delineating the ore body. That one 'find' may not be located conveniently for wildlife or scenic spots. From the company's perspective, the one 'find' must pay back for the cost of all the 'misses.'[44]

Land access, then, is the key to long-term sustainable mining in Canada. If the industry cannot renew its depleting mineral reserves through new exploration, the mining sector will inevitably decline.

One issue that cast the debate into sharp relief was a confrontation over a potential development in northwestern British Columbia. Geddes Resources Inc. wished to develop an enormous copper (with by-product cobalt, silver, and gold) deposit located in the Haines Triangle of British

Columbia, 190 kilometres southwest of Whitehorse, Yukon. The Windy Craggy Mountain, the site of the proposed mine, is located near the confluence of the Tatshenshini and Alsek Rivers, at the headwater of Tats Creek, a tributary of the Tatshenshini. Geddes Resources claimed that the Windy Craggy deposit, if developed, could produce over 165 million tonnes of ore over fifty years. The company stated that it could limit the impact on the environment. After consideration of various alternatives, the company ultimately suggested that it could limit traffic on the roadway running through the river valley by flying in its staff for rotational work shifts. Fuel and the mined copper would be transported as a slurry through pipeline rather than by truck. Waste rock would be placed out of sight in an acid rock drainage lake (formed between two dams and designed to withstand earthquakes). With the exhaustion of ore reserves and subsequent closure of other Canadian copper mines, the industry and resource communities were looking for new mines such as Windy Craggy to fill the gap.[45]

On the other hand, environmental groups (as well as tourism-based operators) were deeply concerned about the ecological implications of development and did not wish to see the deposit developed. Environmental groups produced their own array of concerns when examining the proposed Windy Craggy development and questioned the reliability of the company's statistics and plans for minimizing environmental damage. In addition to their reservations about serious hazards associated with acid-mine drainage, opponents to the project voiced their concerns about the earthquake safety of the tailings dams, the possible impact on the largest eagle sanctuary in the world (the Chilkat Bald Eagle Reserve), and the grizzly bear population. The area is surrounded by American and Canadian national parks – Glacier Bay and Wrangell-St. Elias in Alaska, and Kluane in the Yukon. Well-informed environmental activists acknowledge the importance of resources to a nation's economy, to the generation of employment, and to regional development. This particular case, they said, was different. One group claimed that the Windy Craggy project was the most 'environmentally hazardous mining project ever proposed in Canada.'[46] It has also been suggested that by allowing the project to go ahead, the mine 'would devastate world-class wildlife populations ... poisoning this wild and incredibly biologically rich place.'[47]

In the Windy Craggy dispute, opponents of the project came from a network of North American environmental groups, including the Western Canada Wilderness Committee, the Sierra Club, the US National Parks Conservation Association, and World Wildlife Fund. The issue had interjurisdictional dimensions that included the Alaskan, Yukon, and federal governments, with environmental implications for national parks and aquatic life in the US and Canada. The president of the Canadian Parks and Wilderness Society pointed out that: 'The Tatshenshini wilderness is unique, and more precious than precious metals ... It's among the densest grizzly bear habitats in the world ... There are bald and golden eagles in enormous abundance. It's scenically astounding with big glaciers coming down to the river at 1,000 feet of elevation – one of the wildest I've ever seen ... No compromise is possible on this open pit mine.'[48] For its part, the proponent, Geddes Resources, submitted its Stage I Environmental and Socioeconomic Impact Assessment Report for the project in January 1990. After a project review conducted by provincial and federal agencies, the provincial Mine Development Steering Committee rejected the plan stating that it did not adequately address concerns related to acid rock drainage or prevention of acid generation.

The company submitted a revised mining plan at the end of 1990. On 2 April 1992, a resolution was introduced in the US House of Representatives stating that the Tatshenshini is of international significance and that the area be referred to the International Joint Commission for review. On 6 April, the British Columbia government announced that before any decisions would be made, a full range of land- and water-use options for the Alsek Tatshenshini areas would have to be considered. If mineral development was among the preferred land uses, then a thorough environmental review (including a formal public hearing) of the Windy Craggy project would resume.[49]

The Windy Craggy Project was also subject to a new Mine Development Assessment Process (MDAP), the Environmental Assessment and Review Process in Canada (EARP), and the National Environmental Policy Act (NEPA) in the United States.[50] In July 1992, it was announced that Stephen Owen, chairman of the BC Commission on Resources and Environment (CORE), would study the Tatshenshini Valley over the next year and make recommendations to government about its future use. The government decided that the ecological value of the area outweighed its potential mineral value and, on that basis, designated it a protected area. It is now a World Heritage Site.

BC's decision on the Windy Craggy project had a deep, adverse impact on the confidence of the mining industry in the government's decision-making processes. From the mineral industry's perspective, the issue went beyond Windy Craggy because the whole Haines Triangle was thought to be potentially rich in mineral wealth. According to a Price Waterhouse report, claims held by twenty companies, including Geddes Resources' Windy Craggy deposit, were expropriated: 'Negotiations were expected to commence in 1993 regarding compensation with holders of mineral claims. However, there has not been any negotiation, not even on a process for review and settlement of such claims. The provincial ombudsman is expected to review the process by which the Government reached its decision.'[51] Compensation for the exploration costs and potential earnings lost is a politically charged issue. In 1992, the government appointed the Schwindt Commission to recommend a course of action to determine how (and if) compensation should be awarded. Yet by 1995, little progress had been made in resolving the issue. The question of ownership and appropriate compensation is complicated. The formal constitutional/legal power of the province regarding resource use is broad.[52] The province is the original owner and the regulator of Crown resources. Much hinges on whether or not a 'government taking is viewed as simply one market risk among many or as something else.'[53] As the situation exists, the government is under no legal obligation to pay any compensation. Nevertheless, in August 1995, Royal Oak Mines, which had a controlling share in Geddes Resources' Windy Craggy deposit, was offered a compensation package. Part of the $138 million package would be used to help Royal Oak develop two new copper-gold mines in northern British Columbia by providing infrastructural support and employment. Of that total, $29 million would be spent in compensation for Windy Craggy and $40 million would be spent in mineral exploration in BC.[54] Whether this one agreement will set a precedent for future compensation for other similar cases remains to be seen.

In May 1992, several major environmental groups in British Columbia announced that they were creating an Environmental Mining Caucus of British Columbia. The caucus would be set up to work with the mining industry to establish land-use strategies. Representatives of the new organization said that few mining projects were as sensitive as the Windy Craggy deposit and most could be made 'environmentally acceptable.'[55]

It should be noted that the mineral industry's ability to explore and develop new deposits is *not* threatened through the necessary establishment of responsible environmental reclamation guidelines or the settling of Aboriginal land claims. Rather, the greatest threat comes from the combination of the following factors: an unpredictable world market, the uncertainty created by changing political and regulatory conditions, a lack of investor confidence in the policy environment, and a lack of access to potentially mineral-rich tracts of land.

Conclusion

The Whitehorse Mining Initiative was an attempt to foster a more supportive public policy climate that would provide a framework for continued sustainable mining in Canada. Members of the industry were particularly anxious to educate other policy communities about the value of the industry to Canada and the challenges the industry was facing (as outlined in this chapter). Industry participants expected that the WMI could play an important role in informing key members of interest groups about mining. What the mining representatives were not as prepared for was the extent of the concerns other groups had about the potential adverse impact of mining. During the WMI process, it became evident that the mutual process of education would be extensive, exhausting, and, more often than one might expect, enlightening.

The creators of the Whitehorse Mining Initiative did not have a clear idea of where this uncharted journey would lead. They knew that they would have a difficult job selling the concept. In addition to addressing the challenge of reconciling competing visions and concerns among themselves, industry proponents of the WMI had to persuade other groups to participate. Other communities of interest included stakeholders who often were unconvinced about the utility of the exercise; or were opposed to mining practices; or were unpersuaded about the overall value of mining. Nonetheless, there were many people who were willing to give this new approach a chance. As the next chapter outlines, each of the participants, although broadly grouped under various interest associations (industry, environment, First Nations, labour, and government), arrived at the WMI meetings with many different goals, agendas, and ideological views.

3
Staking a Claim: Broadening the Public Interest

Several different communities of interest participated in the Whitehorse Mining Initiative, including members of industry, mining communities, Aboriginal communities, environment, labour, and government. Representatives of these interests came to the WMI with distinctive policy agendas. The industry clearly wants to foster an environment that would improve its competitive position and is looking for a supportive and coherent public policy environment to sustain mining in the long term. Resource communities need assistance in establishing some buffers against the vagaries of the international marketplace and domestic policies. Aboriginal communities, whose values are so closely tied to the land and who have unresolved land claims, are seeking mining policies that respect their cultural, environmental, economic, and community needs. Environmental groups want to see land use, environmental assessment, and monitoring predicated on principles of ecosystem planning – achieved in part through an integrated land-use planning process and through a network of protected areas. Labour unions primarily focus on issues relating to health and safety, training, job security, and on ensuring that workers will have the ability to organize if they so choose. Government officials see themselves as 'brokers' of the broader public interest, and as representatives of their own governments and ministries.

Each of these constituencies presents a variety of concerns, some in opposition and some complementary. It should be noted that not all members of these diverse policy communities necessarily wished to support or participate in the WMI; some individuals participated reluctantly because they felt pressured to do so. Others recognized that compromise and conciliation with other policy communities was necessary if they wished to influence political agendas. To understand these agendas, it is worthwhile exploring the various perspectives brought to the WMI negotiating table.

The Mineral Industry

The success of a multi-stakeholder approach such as the Whitehorse Mining Initiative is dependent on a number of variables; one of the most important is the readiness of the industry to buy into the process. Roundtable negotiation relies on the willingness of the players to work on a consensus basis within a new public policy environment. This may be a difficult sell, particularly for an industry that can boast considerable technical expertise in activities related to mining, but that lacks political sophistication. As a group, the mineral industry's knowledge of how to get the ore out of the ground far surpasses its ability to develop political alliances and form strategic compromises. The industry is not well adapted to the changing political times, nor do mining interests have much patience for learning to understand the policy process. This characteristic, partly rooted in history and partly in ideology, appears to have changed very little over the past century. For example, in 1932 a group of independent-minded prospectors and developers met to form an organization to protect themselves against what they saw as 'excessive regulations exercised by the provincial bureaucrats against honest prospectors and mining entrepreneurs, and the inadequate governmental protection available to the industry and the public against those who might defraud them.'[1]

This does not mean that the interests of the nation are unconsidered in the quest for opening up new mine developments. This was clear as early as 1933 when Walter E. Segsworth, who served as secretary of the Prospectors and Developers Association, wrote a paper describing the objectives of the association. Among other things he stated, 'It must be recognized that the good of the nation comes first. Also, that any measure we advocate must stand that test. We would lose our influence very quickly if we advocated anything which, while affecting favourably our own selfish interests, was against the best interests of the people as a whole.'[2] As history has demonstrated, however, interpretations about what is in the 'best interests' of the nation will vary and depend on the public values held at a particular point in time.

Members of the mineral industry, prospectors in particular, often share an ideology based on values that promote free enterprise, individualism, and perseverance. At one time, it was enough to excel in exploration and mineral production. In the current political environment, however, industry representatives are finding that they must also become experts on selling other stakeholders on the value of mining itself. But mining representatives are not people who readily embrace ideas of consensus-based decision-making and roundtable discussions. The consuming passion of this enterprise leaves one with little time to consider the broader political and social environment in which such activities take place. It is even more difficult when such discussions must take place with other groups who may not appreciate or publicly acknowledge the personal effort that goes into discovering and developing a single mine.

The mining industry cannot be viewed as a monolithic enterprise. It is tremendously diverse, requiring a wide range of skills and levels of expertise on the part of those engaged in activities such as exploration, development, speculation, processing, etc. Yet, it is not uncommon to hear members of the public perceive a mining businessperson as falling in the same camp as the unscrupulous 'boiler room' stock promoter. This depiction comes as a surprise to many involved in the mineral industry, where mining has been a part of their lives, and that of their families and the generations before them, as is the case with farmers, fishers, and foresters. With mining as an intrinsic part of their histories and lives, many are acutely aware of how they have contributed to the fabric of Canadian culture, society, and the economy. They are proud of their individual initiative, technical sophistication and expertise, their ability to generate wealth for the country, and their work ethic. They are often genuinely bewildered that the general public does not appreciate the role mining has played in Canada.

Mining representatives often believe that the way to deal with the changing political arena is a matter of public education. If only members of the public could understand what mining was all about, the argument goes, they would appreciate its contribution. What is harder for some of those associated with the mineral industry to accept is that no amount of public education will change the fact that many people simply hold different world views about how natural resources should be managed. Given the liberal, individualist ideology of the industry that allowed them to achieve their world-class mining operations, it will be difficult for them to adapt to a political world that requires a whole new set of skills and attitudes.

Industry and the Whitehorse Mining Initiative

The industry came into the Whitehorse Mining Initiative facing competing perspectives from different stakeholders in the external policy community and internally within the mining sector itself. It was unclear to them where the initiative would lead. There were a number of issues that they wanted to see addressed in the meetings. Many of these concerns were discussed in the previous chapter. Some of the major issues were related to restricted land access, the question of compensation for cancelled mineral tenure, the complex, inefficient, and inhospitable taxation regime, overlapping and duplication of environmental regulations, and rising uncertainties in mineral title and environmental liability that make it increasingly difficult to raise risk capital.

Mining representatives entered the WMI discussions warily. What kind of policy and implementation framework should be devised to implement the goals of the WMI? Who will be held accountable for this new process?

Industry remains sceptical of open-ended approaches where an indefinite number of concerned stakeholders may participate in land-use decisions. There is also a larger question posed by the Whitehorse Mining Initiative; that is, does compromise mean that roundtable processes will lead to more or less certainty in the mineral industry? The doubt about the fruitfulness of consensus-based processes is evident in a recent BC mining association publication. Titled *'Leadership' by Consensus? Something to Think About,* the newsletter includes a quotation by former British prime minister Margaret Thatcher: 'To me consensus seems to be: the process of abandoning all beliefs, principles, values and policies in search of something in which no one believes, but to which no one objects; the process of avoiding the very issues that have to be solved, merely because you cannot get agreement on the way ahead. What great cause would have been fought and won under the banner, "I stand for consensus?"'[3] A statement of this kind, quoted in a mining association newspaper, underlines the impression that industry members may not adapt readily to the realities of the modern policy-making and developing cooperative strategies with other communities of interests. Industry competitiveness in today's policy environment requires the adoption of new approaches and attitudes. If the industry can learn to accommodate its needs to the new political marketplace, the WMI could serve as a useful vehicle to achieve the objective of a long-term sustainable mineral sector.

Mining Communities

At the Whitehorse Mining Initiative, mining communities were indirectly represented by a variety of participants, many of whom live in northern or remote towns. In Canada, there are over 150 communities that are partly or completely dependent on mining. While only 88,033[4] people are directly employed in the mining industry, others rely on the job opportunities and economic wealth that is generated by mining activities. It should also be noted that all cities and towns are indirectly affected by mining activities. One need only consider smaller quarrying operations that take place on the rural-urban fringes of communities. For our purposes here, however, we will focus on those communities whose economic health directly relies on mining.

Mining towns are well known for their limited life span. Opportunities for growth and employment have often been determined by factors beyond the control of the community, such as the latest downswing of the 'boom or bust' resource cycle caused by indifferent global markets; depleting ore reserves; the opening of a new 'fly-in' mine where mine workers are flown in from more distant centres; employment cut-backs with the introduction of technological advances; or strategies adopted by the mining companies competing in an international environment.

Towns with primary industry for their economic base have historically been the subject of decisions made elsewhere. The mayor of Kirkland Lake, a mining town in Northern Ontario, observed:

> Kirkland Lake, like many other communities in Northern Ontario, has relied on the natural-resource sector, particularly mining and forestry, for its economic base. Unfortunately, these sectors are often sensitive to the whims of large-city corporate executives, fluctuations in senior government policies, designation, categorization, resource depletion, declining prices, uncompetitiveness and other variables which have spelled doom for many northern communities.[5]

Those well acquainted with mining are all too familiar with the old saw, 'The day a mine opens, it begins to close.' If a new ore body to replace a depleted one is not discovered in a mining area, towns can, and do, collapse. A relatively recent international trend toward long-distance commuting (LDC) or 'fly-in' mining is hastening the demise of existing single-industry towns. Companies seeking to develop a deposit in a remote area quickly realize the economic benefit of flying their workers into a mining site from larger regional centres for days or weeks at a time. The infrastructural and service costs associated with building and maintaining a mining town as well as the attendant environmental costs can be reduced. Residents of smaller resource towns are concerned that the opportunities for employment will 'fly over' them to those larger centres. As for the latter, the more diversified northern cities with good airport facilities can now serve as commuting centres for many fly-in operations.[6] Those communities that do survive will have embarked on a number of community development initiatives to ensure that the community is able to take advantage of different opportunities. Tumbler Ridge, British Columbia, for example, is a mining community built in the early 1980s in the northeastern part of the province for the purpose of servicing a $3 billion project that sells metallurgical coal to Japan. Tumbler Ridge is one of the most recent, and some say one of the last, mining communities to be built in Canada. The modern picturesque community, with its independently elected town council, defies traditional stereotypes and could pass for a vacation resort area with its golf course, community centre, and recreational activities. Although its economy is primarily dependent on two large mines, Tumbler Ridge is diversifying with the development of a gas-processing plant and tourism.

A number of federal and provincial assistance programs are often established to help communities and employees adjust to mine closure. Such programs may include community development programs, mobility assistance to help laid-off miners relocate, training or retraining programs, industrial adjustment services to help employees and workers adjust to workplace changes, and specific community assistance programs. Local governments facing mine closure may have incurred a long-term debt to provide services to a mine-dependent community. They have invested in assets in the areas of housing and other privately held properties, as well as in fixed infrastructure. A few notable programs have been directed at these problems. In 1986, for example, the Manitoba government provided a $55,000 grant from the Mining Community Reserve Fund to the local government district of Lynn Lake. The fund was to help cover the shortfall in municipal taxes, caused in part by the closure of a local mine.[7] In 1992, a Natural Resource Community Fund was established in British Columbia that was designed to provide emergency assistance to resource communities as a result of resource downsizing or closures. One-half of 1 per cent of the province's resource revenues each year will go into the fund. The fund received an initial $15 million that was to be capped at $25 million. It was also to be used to complement other federal provincial programs directed at job training, job creation, maintenance, worker relocation, etc.[8]

In the broader context, however, local governments are going to suffer from the reductions in overall program funding around the country as governments engage in deficit-cutting exercises. Increasingly, senior governments will see themselves in a facilitating role, by providing information to the communities rather than funding. Community development is becoming the 'byword' for community survival in the 1990s.[9]

Community development initiatives may provide some practical economic alternatives, but they can do little to change some of the broader political and economic factors that are affecting resource towns. World market prices are one important variable. Another is the domestic policy environment. For example, northern communities are concerned that lack of public recognition of the importance of mining to the country is resulting in unsupportive policies that lead to a loss of investor confidence. The inevitable result would be a decline of employment in northern communities. Prospectors and developers, labour unions, and mining communities are actively pursuing a variety of strategies to ensure the survival of mining towns. Northern communities engage in public relations initiatives and lobbying efforts to raise their national profile.

'Save Our North,' for example, was a grassroots movement formed in 1991 to alert the general public to the problems facing Northern Ontario communities. With an initially modest media campaign that quickly heated up, the Porcupine Prospectors and Developers Association (PPDA) mustered considerable support from sixty mining communities.[10] This initiative soon gained some attention from provincial politicians. In 1993, a 'Keep Mining in Canada' campaign was inaugurated with the participation of 110 communities. This new initiative was also prompted by concerns about the growing departure of Canadian mining investment to other parts of the world.

In British Columbia, another grassroots organization, Share BC, was formed. As Al Beix, chairman of Share BC, noted: 'The world is run by those who show up.'[11] Local citizens in twenty-four resource-dependent communities throughout the province formed associations to protect their livelihood. Concerned about land-use decisions that may threaten their ability to harvest and extract resources, these associations are attempting to influence public agendas by raising awareness and support.

Communities and the Whitehorse Mining Initiative

Mining communities are undergoing tremendous changes that lead to instability and uncertainty. Although federal and provincial governments appear to endorse greater decision-making and autonomy for local governments, the evidence is unpersuasive that any actual power has been transferred to the local level of government. Community resource boards that provide advice on land-use decisions to provincial governments do not necessarily give them any more of a voice. In fact, on such boards where local governments are only one of many players providing advice to the province, local roundtables could further undermine the ability of municipal governments to represent their communities. We hear a great deal about 'partnerships' between provincial or federal governments and local communities. Partnership implies an equal power relationship between the two groups. This is not the case where local governments fall under the statutory control of the provinces.

There were a number of unresolved issues regarding mining communities that were brought to the WMI negotiations. What kind of adjustment assistance will be provided to communities that are primarily dependent on mines when mines close? There is a world-wide trend toward long-distance commuting to remote mining operations. Will companies use declining mining communities as residential centres for mine workers and their families, or will those communities simply be 'flown over' with employees living within larger regional centres? Does the increasing rhetoric around community development signal a willingness on the part of senior governments to give local communities more authority in determining the future direction of their towns? If it does, will there be an economic cost for that independence? Underpinning these issues was a general recognition that a diversified community is more likely sustainable in the long term. As Jane Jacob notes:

> Economies producing diversely and amply for their own people and producers, as well as for others, are better off than specialized economies like those of supply, clearance and transplant regions. In a natural ecology, the more diversity there is, the more flexibility, too, because of what ecologists call its greater numbers of 'homeostatic feedback loops,' meaning that it

> includes greater numbers of feedback controls for automatic self-correction. It is same with our economies.[12]

It is one thing to know what constitutes a healthy, thriving community. Achieving this objective is quite another undertaking. Representatives of the mining communities came to the WMI to seek support for a policy environment that would allow their communities to flourish and diversify.

Labour Organizations

Canada has been a world leader in mining for a number of reasons, including its rich, natural endowment and historical government policies that favoured resource development, and its effectiveness and efficiency in discovering and producing minerals. With the downturn in the mining industry, and the economy in general in the 1980s, it was the expertise in the mineral sector that increased productivity levels and helped the industry recover. It has been noted that, 'Canadian miners are among the best in the world, respected for finding low-cost solutions to keeping Canadian operations open, and for overcoming some of the most severe climatic conditions. This expertise may be a key to the future of the Canadian mining industry.'[13] A productive, skilled workforce in the mineral industry, therefore, is very important to its long-term survival. Unfortunately, the 1980s restructuring included the adoption of new technologies and methods that inevitably led to a decline in employment levels. Furthermore, a trend in the late 1980s and early 1990s saw the number of mine closures outstrip openings, further reducing employment. Although the outlook has recently improved, many Canadian mines have a short life expectancy with the depletion of reserves. Several thousand employees will be affected by these changes. Employment predictions for mining or mining-related employment in Canada are expected to continue to decrease overall.

The introduction of technological change will also affect the nature of employment. Computerized open-pit mining systems with intelligent supervisory control systems will not only reduce the number of employees required but will call for those individuals with skills in applied science and computer operations. More stringent environmental requirements, for example, require such things as the adoption of purification technologies, and thus workers with expertise in environmental and metallurgical sciences will be needed.[14] Overall, the mineral industry will need more highly educated people with a broad set of skills who can adapt to technological change. These requirements will, of course, vary with the occupation. The exploration sector, for example, will not need the same level of literacy or math skills as those areas that are particularly high tech. Nevertheless, the composition of the workforce will change: 'The current workforce is a mixture of labourers, semi-skilled operators, tradespeople, technicians, technologists, professionals, and managers. The ratio of operators to tradespeople and technicians, traditionally weighted in favour of the operators, is shifting as fewer traditional operators are required'[15] (see Table 3.1). The composition of the workforce is predominantly 'white,' male, and aging. There is relatively low turnover. As a result, there may be limited employment opportunities for younger workers, women, and Aboriginal people.[16]

Organized labour in mining has had to adapt to the changes in human resource requirements, the highly competitive international environment, the uncertain domestic environment, and organizational change. The concerns of unions range from those that are fairly narrowly oriented toward issues related to job security and health and safety, to a more general philosophy of social unionism (part of a broader social movement).

In the case of mining, the agenda of labour organizations may overlap with those of resource communities. They, too, are concerned about the impact of automation, closure, and the trend toward 'fly-in' mining on existing mining towns and their inhabitants. In addition to those concerns mentioned in the previous section, however, labour has other issues with which to contend.

According to Richard Chaykowski and Anil Verma, employee reductions resulting from organizational change and the introduction of new technology have led to a significant decline in union membership.[17] Private sector unions have responded in a number of ways by shifting their focus toward employment security and skills training. There is a growing recognition of the need for union-management cooperation in implementing necessary reforms and innovations in industry to improve competitiveness.[18] Skills training, particularly multi-skilling, has been the subject of much debate. Multi-skilling or job amalgamation has traditionally been resisted by unions based on concerns that the adoption of such approaches by management has not been in the best interests of workers or unions. The end result of many of the new approaches leads to job reductions, the erosion of union-negotiated job-evaluation systems, and inadequate compensation for the acquisition of new skills. Thomas Reid, of the United Steelworkers of America (USWA), suggests, however, that multi-skilling could work if pursued in a climate of trust and cooperation and if an appropriate culture can be created:

> Multi-skilling will win acceptance from production workers, even in very traditionally structured workplaces, if it is pursued cooperatively, with full compensation for the acquired skills and new responsibilities, preferably within the context of a collective agreement and a union-negotiated job-evaluation scheme. Training costs must not be borne by the employee, except insofar as some form of jointly operated training program may already be in effect in the particular workplace.[19]

Table 3.1

Occupational groups in the mining industry

Exploration	Mine	Mill/Concentrator	Smelter/Refinery
Field work/operations (e.g., prospector, surveyor, diamond driller)	**Operations/production** (e.g., blast-hole driller, scooptram operator, development miner)	**Operations/production** (e.g., flotation operator, crusher operator, control room operator)	**Operations/production** (e.g., roaster operator, converter operator)
Trades (e.g., electrician, heavy duty mechanic)	**Trades** (e.g., electrician, heavy duty mechanic)	**Trades** (e.g., welder, millwright)	**Trades** (e.g., welder, millwright)
Technical (e.g., geological technician, assayer)	**Technical** (e.g., ventilation technician, mining technician)	**Technical** (e.g., metallurgical technician, instrumentation technician)	**Technical** (e.g., metallurgical technician, instrumentation technician)
	Supervision (e.g., shift boss, general foreman)	**Supervision** (e.g., shift boss, general foreman)	**Supervision** (e.g., shift boss, general foreman)

Professional (e.g., geophysical technologist, geologist, geophysicist, geochemist)	**Professional** (e.g., mining technologist, mining engineer, mechanical engineer)	**Professional** (e.g., metallurgical technologist, metallurgical engineer)	**Professional** (e.g., environmentalist, metallurgical technologist, metallurgical engineer)
Managerial (e.g., senior geologist, project manager)	**Managerial** (e.g., mine superintendent, mine manager)	**Managerial** (e.g., mill superintendent)	**Managerial** (e.g., smelter superintendent)

Source: Price Waterhouse, *Breaking New Ground: Human Resource Challenges and Opportunities in the Canadian Mining Industry* (Ottawa: Ministry of Supply and Services Canada 1993), 32.

It is particularly difficult in the mineral sector to develop continuity in labour relations because of the tremendous variation in the workplace depending on the size, scale, type, and location of an operation. As Chaykowski notes, working conditions will vary from small-scale, short-lived, fly-in or remote operations, to large-scale enterprises (including mining, smelting, and refining) that have sustained sizeable mining communities such as Sudbury, Ontario, for many years. The nature of the enterprise will very much influence the ability of the operation to attract and retain a skilled workforce. It will also affect management-labour relations. These characteristics will have an impact on the ability of unions to organize workers.[20]

Yet Ken Delaney of the United Steelworkers claims that the relationship between labour and management could be more cooperative in ways that would assist both the mining companies and workers. He suggests that different mining approaches could be employed, 'approaches that extend the life of the mines and the jobs of the workers (e.g., avoiding 'high-grading' approaches to mines). Improvements in quality and productivity should not be at the cost of labour's health or by increasing workers' stress levels. This is not in the long-term interests of the organization or labour. Yet this is what has been happening.'[21]

In sum, changes in the international markets and technology have led to such developments as workplace automation and fly-in remote operations, which are having a significant impact on the goals of organized labour. Labour organizations are now focusing their agendas on helping workers adjust to the uncertainties of employment in the 1990s. To achieve these goals, labour would benefit from a consensus-based policy approval such as the WMI.

Labour and the Whitehorse Mining Initiative

Labour representatives came to the Whitehorse Mining Initiative with a variety of concerns. Among other things, they were looking for ways to achieve a measure of employment stability for the unions and their membership, in an unpredictable and diverse industry. Specific objectives included maintaining and enhancing training opportunities for the workforce to ensure that they could develop useful and portable skills (that could be applicable both inside and outside the mining industry). Other issues that were targeted included the lack of national occupational standards for professionals, trades and other vocations, as well as a lack of national standards in the area of health and safety regulations.[22] More broadly, there were other goals that were not issue specific. For example, some labour unions see themselves as a vital component of a democratic and just society. It has been argued that unions are important to the long-term interests of Canada's economy and society:

> Unions ... have a critical role to play in countering an increasing polarization of society between an increasingly well-off minority and an increasingly disadvantaged and disillusioned majority. This polarization is not only morally reprehensible; it is almost certain to lead to increasing social unrest, to everyone's disadvantage. Social and industrial conflict is one of the greatest enemies of economic performance.[23]

Perhaps it is in this context that labour sees its largest contribution to the Whitehorse Mining Initiative. Moreover, if the mineral industry is to be sustained, it will need to work with unions to achieve an efficient and productive workplace. For their part, the unions need to contribute to such roundtable initiatives if they wish to help set the agenda of mining policy in the years to come.

Aboriginal Peoples[24]

Aboriginal peoples have many different cultures and languages but they share a deeply rooted connection with the natural environment on which they have a historic dependence. The interdependency among people, and between people and the environment, was vital to their survival.[25] This holistic approach to nature, where people are viewed as a part of the web of life, has shaped their values and world views:

> The work ahead is as everlasting and stochastic as the natural world. Our priorities in natural resource management must shift from a linear model to a holistic approach. To continue performing our duties, as human beings were meant to, we must possess: a proper attitude toward Mother Earth; a respect for cultural differentiation; the proper resources to accomplish set goals; the right tool to overcome future difficulties; a keener knowledge to synthesize the components of ecological and social systems; and, to accomplish this task, a strong spiritual and intellectual connection with the natural world.[26]

Such a perspective fits well with the complex demands of a natural ecosystem and a philosophical orientation that sees human activity contained within a particular bioregion. Yet modern industrial societies, particularly those such as Canada that are economically dependent on the export of natural resources, are not predicated upon similar values. As Carylin Behn points out, however, Native communities are adapting to industrial society and 'evolution is necessary for the survival of any culture.'[27]

In the resource-based relationship initiated over 500 years ago between Aboriginal peoples and Europe, Native peoples played a very important role in resource development. They served as guides and shared their knowledge of the land and their survival skills. By the twentieth century, Aboriginal peoples were relegated to the peripheries of resource development and land-use decisions. At the end of this century, however, some changes are starting to take place. Aboriginal peoples are now beginning to play an important part in natural resource decision-making.[28] Daniel Johnson identifies some of the attempts to include them in resource development:

> Many jurisdictions now require resource companies to enter into Human Resource Development Agreements as a condition of obtaining land leases. In northern areas, this inevitably brings resource interests into working relationships with aboriginal communities. Recently concluded and pending aboriginal land claims have given these groups surface and sub-surface rights or significant tracts of land in addition to guaranteed participation on land management boards. Claims have also resulted in the payment of large sums of money to aboriginal groups, funds which may be used for venture capital purposes.[29]

In spite of these few examples, however, there are serious barriers to resolving divergent perspectives concerning land claims and self-government. Funds may be used for venture capital, yet Aboriginal communities in general often do not have the human resources (e.g., knowledge of developing a business, bookkeeping skills, etc.) to maximize the benefit. Such skills are often provided by outside consultants. People not only need to have the resource, they need to have full knowledge about how to make use of it. Furthermore, it should be noted that the opportunities for resource development in the south may be more limited than those in the far north, given, for example, demographic differences.[30] Training, education, and economic assistance has not been broadly available in ways that would enable Aboriginal peoples to become self-sufficient and equal partners in resource development. For this to happen much depends on the outcome of treaty negotiations and self-government initiatives, accompanied by sufficient economic assistance that would allow indigenous peoples to regain their independence. Behn suggests that:

> If there were adequate structures in place within existing frameworks, attaining self-sufficiency would not be dependent on the outcomes of treaty. However, because non-aboriginal society had policies of exclusion for so long, native people now hope treaties will provide the resources necessary to remove native people from the poverty cycle with all its intrinsic ills. Historically, though native people had less materially they were not 'poor.' The way I see it our greatest loss was the ability to sustain ourselves and our families – that is, to provide a living for ourselves and loved ones. We had the skills to do this in a land based culture but are only now acquiring the skills to do so in a wage based culture.[31]

One cannot discuss Aboriginal issues, however, without recognizing the historic and central importance of land claims and treaty negotiations to the political position and social needs of Aboriginal peoples. They state that they have continued to occupy and use their lands since it was bestowed upon them by the Creator. Their claims to land title, the quest for 'self-government' and 'self-determination,' and their claim as the First Peoples of Canada serve as a common link for the many diverse tribal groupings in Canada. As political scientist Paul Tennant points out, 'underlying the land claims ... is the passionate Indian desire that non-Indians ... acknowledge and appreciate the simple historical fact that Indians were present in established societies of high attainment before Europeans arrived.'[32] Furthermore, self-government has been defined in a way that protects the special status of indigenous peoples and recognizes the independent status of Aboriginal peoples – an awareness that another order of government be established.[33] Despite the many different nations that exist in Canada, the overall position articulated by many Aboriginal leaders is broadly consistent. They state that the jurisdictional powers of their governments are:

- inherent, having their origin in acts of creation
- based upon traditional laws
- rooted in a very special relationship with the land and resources
- constitutional as well as functional – that is, they include the right to determine, for example, basic political constitutions and requirements for citizenship, as well as to provide for the social, economic, cultural, and political well-being, in the broadest sense, of their citizens
- fiscal, including the right to control external relations concerning land and resources, as well as to tax and collect revenues
- independent of the jurisdiction of Canadian governments and protected by international law, as well as convention
- assumed and asserted as these governments perceived a need to do so.[34]

Before 1929, eighty treaty agreements were signed with Aboriginal peoples. In 1973, the federal government accepted that it needed to negotiate 'comprehensive' land claims settlements where Aboriginal title had not been extinguished by treaty. Under these modern treaties, several agreements followed, including the James Bay and Northern Quebec Agreement of 1975 and the Northeastern Quebec Agreement in 1978, the Inuvialuit in 1984 (parts of the western Arctic in the Northwest Territories and the Yukon Territory), the Guich'in in 1992 (lower Mackenzie Valley), the Tungavik Federation of Nunavut in 1993 (eastern and northern Arctic), and the Council of Yukon Indians in 1993. In British Columbia, in December 1991, the provincial government recognized the inherent right of Aboriginal peoples to self-government. In 1993, the BC legislature passed the Treaty Commission Act to initiate a process for the negotiation of treaties. The BC treaty negotiations are now underway. On 16 February 1996, the first modern Indian land-claim treaty was launched with an agreement-in-principle signed by federal and provincial governments and Nisga'a representatives. The Nisga'a people reside in northwestern BC in the Nass Valley near the Alaska border. Under the interim agreement, the Nisga'a are to receive $201 million and to have communal ownership of 1,950 square kilometres of their traditional territory. Furthermore, the agreement granted the rights of the Nisga'a people in certain areas of commercial fishing and forest tenure, and to pass some laws regarding culture and language, employment, traffic, policing, and land use, in conformity with federal and provincial standards. The Nisga'a would begin paying sales tax in eight years and income tax in twelve years.[35]

Constitutionally and legally, Aboriginal rights have been recognized. Section 35 of the Constitution Act, 1982, recognizes and affirms the existing Aboriginal and treaty rights of the Aboriginal peoples of Canada. Section 25 of the Charter of Rights and Freedoms guarantees that 'certain rights and freedoms shall not be construed so as to abrogate or derogate from any aboriginal, treaty or other rights or freedoms that pertain to the aboriginal peoples of Canada.' Increasingly, Aboriginal rights are being recognized in law. For example, one particular case signals an increasing legal recognition of Aboriginal rights vis-à-vis natural resources. In the 1990 *R.* v. *Sparrow* Supreme Court of Canada decision, the court unanimously ruled that Aboriginal peoples have a collective right to fish for food, as well as to their ceremonial and societal purposes. This takes precedence over other rights of use. This has been interpreted to mean that provincial laws and regulations must be drafted in a way that recognizes and affirms Aboriginal rights, with minimum interference.[36]

A second case is that of the 1991 *Delgamuukw* v. *Her Majesty the Queen in Right of British Columbia and The Attorney General of Canada* and its subsequent appeal in 1993. The BC Court of Appeal reversed the lower court decision ruling that Aboriginal rights in BC have not been extinguished.

Legal challenges, treaties and constitutional recognitions, and general awareness about the important role that Aboriginal peoples have to play in resource decisions have enhanced the role of Aboriginal peoples in resources decision-making. In British Columbia, for example, the Ministry of Energy, Mines and Petroleum Resources developed several policies with respect to Aboriginal rights in resource development. These policies are based on the spirit of principles of recognition, cooperation, and respect in order to develop a good working relationship with Native communities. Consultations will be conducted in a similar manner as those held on a government-to-government basis; that is, in a way that will foster joint cooperation in the management of land and natural resources, and in a way that appreciates Aboriginal cultures, traditions, and values.[37] With respect to Crown land, the ministry 'will give priority to an aboriginal right where energy and mineral resource activities demonstrably limit, impede or deny the exercise of that right.'[38] Those rights shall not be infringed unless it is 'unavoidable,' as in the case of a resource conservation or management need, public safety, or another compelling objective.[39] Consultations with Aboriginal peoples are expected to take place at all levels of impact when there is significant mine exploration and development.

There are a number of ways that Aboriginal peoples are seeking to extend their responsibilities and rights in the area of land and resource management. Aboriginal communities require the tools that will allow them to re-establish many of their traditions in a way that harmonizes with modern lifestyles. With the prospect of cash settlements that accompany land claims, Aboriginal peoples can participate in Native organizations and businesses that are organized at the band level to create a locally controlled economy. They may be in a position to develop the resources through partnerships, or in cooperation with the mineral industry, and exert more control over resource decisions that have a fundamental impact on their lives.[40]

Aboriginal Peoples and the Whitehorse Mining Initiative

Aboriginal peoples would like the opportunity to participate in resource decision-making regarding mining, either to benefit from mineral development or to help establish policies governing the wise use of the resources.

Robert F. Keith, as a long-standing member of the Canadian Arctic Resources Committee, conducted a series of discussions with Aboriginal peoples across northern Canada in the fall of 1995. In so doing he identified a number of issues, themes, and opportunities for Aboriginal peoples related to mining. They included the following concerns: First, it was noted that the industry and government need to address the lack of notification, consultation, and consent of Aboriginal communities during exploration and mining following established legal processes. Regulations could also be revised to address the environmental and sociocultural impacts of exploration. These regulations could be informed by traditional knowledge and experience of the regions. Traditional knowledge, in conjunction with the collection of scientific data and information, can and does play an important role in the conducting of regional baseline studies. Impact and benefit agreements could be negotiated in order for mining to bring jobs and other business opportunities. Training should go both ways. Aboriginal peoples would like to learn more about mineral exploration, development, and operations in order to benefit from possible opportunities. On the other hand, members of the mining industry could benefit from learning more about Aboriginal cultures, traditions, and economy in order to minimize adverse impacts. Sociocultural community planning should be part of overall mineral development initiatives. Environmental protection is considered extremely important in all aspects, including protecting wildlife, habitats, and important harvesting areas. Environmental impact assessment processes need to be reviewed, in particular, to clarify roles and the harmonization of responsibilities. There is a need for integrated land-use and resource-planning regimes informed by good regional baseline studies, thus reducing ad hoc responses to mining development proposals. Other economic interests must also be recognized through impact agreements and improved use of existing knowledge and revised regulations. These interests include traditional economies and wildlife harvesting, as well as tourism development.[41]

There are a number of different agreements that are in place where land claims remain unsettled throughout Canada. They include specific joint stewardship (or co-management) and interim measure agreements. Interim measures are time-limited and negotiated outside the treaty process, and are not binding in the treaty negotiations. These interim agreements are not without their difficulties. Native peoples are not assured that their interests and concerns regarding resource use in the land claims area will be satisfactorily addressed in the interim or in a final agreement. They are also concerned that such agreements could jeopardize the claims process. For their part, mining companies are not provided with the levels of assurance they would like concerning the completion of their mining projects. Companies are also unclear about lines of accountability, the parameters for consultation, the nature of the regulatory structure, and final authority regarding these arrangements.[42]

Historically, the economic and employment opportunities available to Aboriginal peoples were in relatively unskilled, low-paying jobs. Accessible training and education programs will be essential if Aboriginal peoples are to participate fully in the mineral industry. Participation in the mineral sector may include anything from prospecting, mapping, surveying, and mining to participating in joint ventures. To be successful, training programs will need to accommodate the sociocultural, economic, and sometimes political realities of the Native community.[43] The proximity of Native communities to mining could also provide opportunities; over 30 per cent are located within fifty kilometres of a mine.[44] There are many ways in which Aboriginal peoples can make a valuable contribution to the resource sector in general. Behn points out that 'the whole area of incorporating indigenous knowledge into resource development is virtually untouched.' She suggests that in terms of environmental issues,

> a community member's knowledge could be helpful because of their knowledge of the region, habitat and wildlife into project construction, reclamation and so on.

> There are ways to adapt traditional skills and knowledge to industry economy in a way that is good for us all. Aboriginal people can utilize these skills, make them marketable and companies can build projects that are more environmentally sensitive.[45]

Wage employment does have its problems, however, particularly in northern remote areas. To achieve employment, Aboriginal peoples often have to move outside their communities to where the jobs are located. Though it does provide employment, commuting mining can also have a disruptive impact on family and community life.

An alternative to Aboriginal participation in mining through the wage economy as an individual worker would be through community development. As yet there are a few, but limited, partnerships between mining companies and Aboriginal communities. Up until the WMI, there had not been any coordinated strategy to develop such partnerships, although some precedent had been set in energy and forestry sectors. Recently, an association of Aboriginal mineral professionals, the Canadian Aboriginal Mineral Associations, was formed to help facilitate Aboriginal participation in mining. The Whitehorse Mining Initiative, which included Aboriginal participation, could have the potential to promote the development of a much-needed coordinated strategy.

Environmental Communities

Burgeoning public concerns about the unsustainability of the world's cumulative activities are having a notable impact on government policies. Environmental perspectives can be roughly categorized into two main approaches – the anthropocentric and the biocentric. The anthropocentric perspective is human-centred and focuses on the interaction between human activity and the environment. Consequences of that activity are only considered in terms of how humans are affected. The biocentric approach is ecology-centred and suggests all creatures are equally important; humans must understand their place within the ecosystem and be able to work with nature rather than subdue it for their own interests. While there are different streams of environmentalism, the concern about the long-term health of the ecosystem is the primary orientation of most environmental activists. 'Ecology' has been defined in the following way:

> A fundamental principle of ecology is that an ecosystem is more than the sum of its individual parts; that is, just as the properties of water are not predictable from the individual properties of oxygen and hydrogen, so the emergent properties of ecosystems are not predictable solely from the properties of the living entities and nonliving matter of which they are

> composed. Each ecosystem on Earth must be understood in terms of the interactions of its components.[46]

Many environmentalists embrace this 'holistic' perspective. Resource development and other human activities must take place in a way that can harmonize with the needs of the ecosystem. This is difficult to achieve in a liberal, capitalist society, where capitalism is predicated on economic growth and liberalism rests on tenets of individual and property rights.

Beyond the environmental problems posed by economic and political forces, it is questionable how many people will adopt in practice the ecological principles that they acknowledge to be important. Many will agree that current global population growth and consumption trends are environmentally unsustainable and acknowledge the need for environmentally sound measures for the sake of the collective good. Yet, individuals are less willing to accept the principle of the collective good if it comes at the cost of relinquishing control over certain private property rights or standards of living. This is the heart of the difficulty of achieving a sustainable society. More fundamentally, there is the question of how to harmonize basically incompatible principles inherent in liberal democratic societies: How should we define and accommodate individual rights while striving for the common good? How can we define the common good in the context of sustainable development?[47]

Bioregionalism is one approach that recognizes that, as in nature, individuals are part of an interdependent community. It combines a land ethic with a communitarian spirit:

> *Bioregions* are the natural locales in which everyone lives. *Reinhabitation* of bioregions, creating adaptive cultures that follow the unique characteristics of climate, watersheds, soils, land forms, and native plants and animals that define these places, is the appropriate direction for a transition from Late Industrial Society ... agricultural and natural resources policies can obviously be linked to restoring and maintaining watersheds, soils, and native plants and animals. Energy sources should be those that are naturally available on a renewable basis in each life place, and both distribution systems and uses for energy should be scaled in ways that don't displace natural systems. Community development in all its aspects from economic activities and housing to social services and transportation should be aimed toward bioregional self-reliance.[48]

The difficulties in achieving the ideals of bioregionalism are many. Information technology is leading toward a greater integration of local, provincial, national, and international boundaries, which fosters an increasing dependence on external forces for the demand and supply of goods. The global economy and markets are predicated upon trans-border flows of goods, capital, and labour and information, and the requirements of huge metropolitan cities depend on political and economic forces taking place far beyond their boundaries. Global economic and political interdependence does not necessarily promote a sense of community. Furthermore, our governing institutions have historically done a reasonable job of entrenching the recognition of individual property rights as a principle of liberal democracy. These institutions have been far less successful at fostering a sense of community responsibility.

Those who are actively trying to raise public awareness about the environment face a similar conundrum to that of the resource industry. Given that members of the public are becoming increasingly removed from the source of their well-being, they have little understanding about how their activities impact on the environment and the economy. Aldo Leopold once wrote:

> Perhaps the most serious obstacle impeding the evolution of a land ethic is the fact that our educational and economic system is headed away from, rather than toward, an intense consciousness of land. Your true modern is separated from the land by many middlemen, and by innumerable physical gadgets. He has no vital relations to it; to him it is the space between cities on which crops grow. Turn him loose for a day on the land, and if the spot does not happen to be a golf links or a 'scenic' area, he is bored stiff.[49]

Little has changed. For the average urban dweller, environmentalism may be what some have referred to as 'selective environmentalism.' This might mean that individuals will occasionally fund environmental causes, sport T-shirts featuring an endangered species, faithfully participate in a bluebox program, join in a rally, or buy products in plastic bottles that purport to be environmentally friendly. What one does not see, however, is widespread sustained public support for initiatives that would radically change patterns of consumption, housing, and lifestyle. Moreover, the appropriate steps to take that would ensure a healthier environment remain unclear. Leopold's prescription about land use was to stop thinking of it solely as a land-use problem: 'Examine each question in terms of what is ethically and aesthetically right, as well as what is economically expedient. A thing is right when it tends to preserve the integrity, stability, and beauty of the biotic community. It is wrong when it tends otherwise.'[50] Leopold, however, was also realistic about the likelihood of change and the need for optimism and persistence. He knew that a change in direction toward a wider 'ecological' conscience would take time. Most important, he said, is that we take the right direction in matters of land use:

> If we grant the premise that an ecological conscience is possible and needed, then its first tenet must be this: economic provocation is no longer a satisfactory excuse for unsocial land-use, (or, to use somewhat stronger words, for ecological atrocities) ... Cease being intimidated by the argument that a right action is impossible because it does not yield maximum profits, or that a wrong action is condoned because it pays. That philosophy is dead in human relations, and its funeral in land relations is overdue.[51]

The problem with pursuing the above goal is that 'ethics' and 'aesthetics' are very much culturally bound values, and subject to numerous interpretations. Economics does not solely determine all land use; politics does as well. When there must be trade-offs between environmental, development, and other public goals, it is very difficult to determine what the trade-off should be. Furthermore, a philosophical holistic perspective predicated on the patterns of the 'natural world' (however interpreted) is not compatible with the goals and demands of modern-day technological society; nor is it suited to large-scale, export-oriented activities such as mining on which such a society has come to depend – a society built on so-called 'frontier economics.' The ecological requirements of the world, then, are not easily reconciled with the imperatives of capital.

Environmental Communities and the Whitehorse Mining Initiative

Those members of the environmental community who entered into the WMI negotiations were looking for an approach that would see the activities of the mineral industry interact in a more harmonious way with the ecosystem. In so doing, some hoped that they might also be able to encourage a stronger set of environmental ethics in the resource industry. As Annie Booth, a professor in environmental studies, points out:

> One problem is that ethics may be imposed from the top down, that is, from the society to the individual. But, to function effectively, an ethic requires individual and personal commitment. It requires a willingness to freely make the personal sacrifices of individual liberty which may be required, and to accept personal responsibility for living within ethical constraints. An ethic which is accepted only because society requires it will be subject to minimal compliance, evasion of personal responsibility, and outright defiance, if the individual or group is sufficiently powerful or arrogant.[52]

The Whitehorse Mining Initiative provided the opportunities to discuss different world views and to educate others about environmental ethical issues. That said, however, environmental spokespersons also came to the WMI with a more specific set of goals.

At the top of the agenda was the goal of completing Canada's protected areas networks and ecologically sensitive areas. Environmental representatives have endorsed the Canadian Wilderness Charter, which emphasizes a need to take steps against environmental threats to the Earth by current human activities that are endangering many species and ecosystems. By December 1993, this charter was signed by over 250 non-governmental organizations.[53] One of the goals of the charter is to promote the development of a network of protected areas that encompasses a full range of ecosystems; this goal was backed up with the Tri-Council Statement of Commitment to Complete Canada's Network of Protected Areas. The commitment was signed in 1992 by the Canadian Council of Ministers of the Environment, the Canadian Parks Ministers' Council, and the Wildlife Ministers' Council of Canada. The statement included the following commitments:

- complete Canada's networks of protected areas representative of Canada's land based natural regions by the year 2000 and accelerate the protection of areas representative of Canada's marine natural regions
- accelerate the identification and protection of Canada's critical wildlife habitat
- adopt frameworks, strategies, and time-frames for the completion of the protected areas networks
- continue to cooperate in the protection of ecosystems, landscapes, and wildlife habitat
- ensure that protected areas are integral components of all sustainable development strategies.[54]

For the mineral sector, these goals would mean that exploration and mining activities would be prohibited in certain places such as in protected areas. In other circumstances, the goal was to work toward sustainable development by encouraging mining activities to take place in an environmentally responsible manner.

By participating in the WMI forum, participants implicitly acknowledged that the goal of protecting the natural environment would have to be pursued within a framework of sustainable mineral development. Those who do believe in the spirit of cooperation and consensus-building recognized that there may be some middle ground between the extremes; a compromise had to be struck between the economic goals of development initiatives and deep ecology, which offers few ready approaches for dealing with the realities of a modern capitalist society. While sustainable development was generally seen as a starting point, it was unclear how that could be effectively reconciled with an ecosystem approach to resources use.

Government

Governments have a number of different goals when it comes to participating in exercises such as the WMI. It is important to recognize that they, like the other interests mentioned above, are not all 'singing from the same songbook,' nor are they necessarily in tune with each other; nor, for that matter, should they be in a liberal-democratic society. Canada's federal-provincial division of powers and the evolution of different political cultures throughout the country have led to a highly diverse political system. Goals will also differ within one government organization and between the elected representatives and the administrative arm of government. Elected representatives need to maintain public confidence to ensure that they are re-elected and must be seen to represent the public interest. Public servants are directly responsible to their political masters through the concept of ministerial responsibility. Beyond that, they are expected to administer, and often interpret, the government policies of the day. Government departments, themselves, will have competing objectives and goals. Differing agencies and ministries will vie for power to maintain their own legitimacy and influence within the bureaucracy. In so doing, they have developed their own political cultures and set of alliances. As a vehicle that attempts to reconcile some of these differences, the Whitehorse Mining Initiative may prove to be the most useful approach to carrying out governments' interpretation of the 'public interest.'

A climate of heightened public expectations has evolved over the past few decades. Public interest groups increasingly expect to be consulted and participate meaningfully in the government decision-making process. Processes that are not perceived to be inclusive and open lose legitimacy in the eyes of vigilant publics. Other factors that have fuelled public demands to participate in decision-making include broader political trends, such as a rapidly eroding vision of a postwar world order; a declining public confidence in the ability of the state to deal with escalating environmental and economic questions; and the quest for economic diversity. With the growth of interest group activity, governments often feel compelled to undertake a broad consultative process in making major decisions about either the natural environment or proposed economic development. Marrying the twin concerns of a healthy environment and a healthy economy will not necessarily lead to a harmonious integrated community of interest.

Governments are now in the position of wrestling with the need to handle urgent environmental, social, and economic issues in a decisive, rational manner. On the other hand, sustainable development cannot be realized without broad public consultation, education, and participation. The inevitable compromise may end up satisfying few.[55] Ironically, the more governments are called on to plan and the more the public is asked to participate, the more complicated the decision-making process. In addition, the need for environmental protection is high while the availability of revenues to take proactive planning strategies is low.

Newly elected governments draw their support from definite constituencies of interest. The longer a particular governing party holds office, however, the wider it must cast its net to maintain support by brokering a number of competing interests. Brokerage politics waters down the ideological underpinnings of a governing party, particularly as the government seeks to extend its legitimacy in the eyes of the majority of the electorate. Therefore, in the formulation and administration of mineral policy, the role of governments has been to regulate the industry while at the same time provide a supportive environment that would allow the industry to prosper. The requirement that governments balance these conflicting goals leaves them in an unenviable position. The same government can be perceived as being 'in the pockets' of big capital, operating at the behest of multi-national corporations, or as the weak 'pawn' of anti-business interest groups, lacking political will and leadership.

Governments have provided considerable support to the mineral industry over Canada's history. Without that support, the country would not be a world leader in mineral exports today. Government assistance to the industry has taken many forms and extends from the international to the local arenas. While jurisdiction of mineral resources in Canada belongs to the provinces, the federal government plays a significant role in many areas, including coordinating regulatory and policy activity, international trade, Aboriginal affairs, science and technology, and in areas where its jurisdiction overlaps with the provinces, such as environmental issues.

Governments offer assistance to the mineral industry at all stages, from exploration to marketing:

> Provincial, territorial and federal governments all have taxpayer-supported mines and minerals departments. They provide a variety of services to the mineral industry, government agencies, and other clients including: geological survey activities, administering land tenure, managing statistics, setting policy, attracting investment, promoting export of commodities and services, and conducting research and development in all phases of the mining cycle ... In a very real sense, government and industry are partners in the development of mineral resources.[56]

An excellent example of this is the federal/provincial Mineral Development Agreements (MDA), which were implemented to stimulate the mining industries. One series of MDAs, which extended from 1984 to 1989, totalled $261 million. Mineral Development Agreements were renewed during 1990-5 but at a reduced amount.[57] The agreements focused on geoscience, extraction and processing technology, economic development, and public information.[58]

At the international level, governments play a role in maintaining a stable international market environment for particular commodities. More generally, one observer explains how the federal government's Mineral Policy Sector's (now the Minerals and Metals Sector) activities affect the interests of the mineral industry:

> MPS works closely with External Affairs in bilateral and multilateral negotiations involving mineral trade, and with foreign officials to facilitate Canadian corporate contacts. Its mineral commodities officers are frequently consulted by the industry, particularly but by no means exclusively, by its marketing people. The mineral industry also depends on the MPS to understand and explain the mineral sector to ensure balanced consideration of its special needs and nature throughout ... government.[59]

The level of assistance governments may be able to provide industry is, however, declining. In 1995, the federal government undertook a widespread budget-cutting exercise in an attempt to reduce the deficit. In the process, the budget of Natural Resources Canada was cut by 50 per cent. The mineral development agreements will not be renewed, resulting in a sizeable reduction in research funding. The availability of technical input from Natural Resources Canada (through the Geological Survey of Canada, CANMET, and the Mineral Policy Sector) has an impact on the industry's ability to compete internationally.[60] Provincial governments are also acting on a desire to reduce government spending. As a result, the mineral industry is concerned about the continued provision of what they term 'essential services,' such as geoscientific knowledge, expertise, and information data bases.[61]

Regulations governing the industry, and regulations that are not directly aimed at mining but that still have an impact, are becoming so complex that industry claims the cumulative costs make it unprofitable to develop a mine in Canada. Regulations can cover issues related to land access and tenure, comprehensive environmental protection (from the protection of a local fish spawning area to the international conventions governing the transboundary movement of hazardous wastes), health and safety, foreign investment, transportation, international trade agreements, and taxation. The mineral industry often feels that governments do more 'to' them than 'for' them. The shifting regulatory environment – as a result of rapid changes in the domestic and global socioeconomic environment, technology, and markets – does not inspire confidence in investors. As governments are asked to distribute a diminishing share of the pie to evergrowing numbers of participants, it is unlikely that resource industries will ever again command the support and attention of governments that they once did. The ability of governments to satisfy the rising number of players will continue to diminish.

Michael Porter's book, *Canada at the Crossroads: The Reality of a New Competitive Environment,* emphasized government's key role in the future competitiveness of the economy. He states: 'The question is not whether government should have a role, but what that role should be ... Government policy should be directed to building the skills, research infrastructure, and other inputs on which all firms draw.'[62] The role of government is not simply to provide a favourable policy environment that assists industry. It must also mediate and regulate competing interests, and ensure that there are sustainable social, as well as economic, benefits from the exploitation of the resources. Room to manoeuvre is limited as the world becomes more interdependent. Externally, the world markets are also responding to a diversity of forces that presents users of raw materials with an array of choices. Markets and governments are not going to adapt to the demands of the mineral industry. A range of strategies is going to have to be employed to accommodate these developments. The Whitehorse Mining Initiative could facilitate that accommodation if it is carefully implemented and if it avoids the pitfalls of many other consensus-based land-use exercises.

Government and the Whitehorse Mining Initiative

Multi-stakeholder approaches such as the WMI are viewed with some trepidation by seasoned bureaucrats. In the first place, many members of the public service were not necessarily trained in the art of political negotiation and compromise. Engineers, biologists, economists, geologists, and other scientific and technical experts now often find a good part of their day occupied with attempts to achieve some sort of compromise between competing interests, rather than getting on with the jobs for which they were trained. Government civil servants are rarely happily engaged in pursuits that result in them being held accountable to an ever-broadening constituency of participants. Moreover, many participants in multi-stakeholder approaches know what they want but not how to achieve it politically. They suggest that it is up to government to find the answers. A further complicating factor is that participants have varying degrees of interest in, or commitment to, these processes.

Participants do not always realize that their demands have to be sold politically. The answer is not simply to replace one governing administration with another that might be more sympathetic. Politically astute interest groups must be able to provide the administration with solutions and approaches that the bureaucrats and their ministers can sell to their colleagues, the media, and the general public. It is not surprising, then, that long-time participants in the political and government scene are aware of the pitfalls that accompany public consultation exercises such as the WMI. Furthermore, members of government must be persuaded that the WMI will lead to solutions and approaches that they were not already implementing and adopting. Like others, if they are to spend their time in endless rounds of negotiation, public servants would like to see some practical and applied results.

Conclusion

The formation of efficient mechanisms for public participation in the policy process, then, is essential if the WMI is to have a long-term impact and not die a slow death. At the policy-making level, geographer Bruce Mitchell suggests that conflict and uncertainty must be dealt with by using approaches that would 'ensure that interests with a legitimate concern have an opportunity to shape the nature of management and development decisions.'[63] While it may be very difficult to discern whose concerns are 'legitimate,' Mitchell suggests a combination of four approaches: Balance, Ecosystems, Adaptiveness, and Teamwork (BEAT). Mitchell emphasizes that to adopt a comprehensive ecosystem approach would result in an 'unrealistically broad scope of the problem.'[64] He, therefore, favours an integrated rather than a comprehensive planning approach. This approach attempts to deal with competing interests while recognizing the need for some overarching policy framework.

The task of integrating the competing vocal interests, however, in the current policy environment is requiring governments to adopt new frameworks of decision-making and public consultation. These new approaches must be perceived as legitimate by both the participants and the attentive voting public. The search for consensus is leading decision-makers down a very different path than has traditionally been the case. Success in these new directions in policy-making is predicated on effective communications, the ability to reach agreement on important policy issues, and public acceptance of these roundtable negotiations as legitimate political avenues for interest accommodation.

4
Rough Terrain, Rich Resource: The Whitehorse Mining Initiative

> The real interest for Canadian mineral policy is how various interests are to be reconciled without merely attempting to average them, and which priorities will be chosen as paramount. Such balances as can be achieved cannot be determined a priori, nor can they be cast in stone. Policy, while it cannot be made according to blueprints and precise long-term planning, must aim at stability and coherence ...
>
> The difficulties are legion, but that is no excuse for an abdication by any of the interested parties.[1]

The WMI process represents a distinctive Canadian evolution in policy-making. It builds on 'the Canadian instinct for consultation and consensus rather than adversarial processes (as are evident in the United States).'[2] As observed in previous chapters, the sociopolitical context of policy-making has changed and there has been a substantial erosion in the ability of governments to balance the contentious claims of an increasing array of interest groups. Traditional equilibrating mechanisms within the policy arena cannot respond effectively to the diverse values of the vocal interests that wish to participate actively in the policy process. Neither the firefighting capabilities of the civil service nor scientific management skills can suppress the tenacious call for 'inclusion' or resolve the complex web of policy issues on the political agenda.

In land- and resource-use policy arenas, attempts are now being made to apply mediation techniques to resolve conflict. These initiatives are based on 'roundtable' approaches in which stakeholders identify and address the facts, debate the issues, and strive to attain resolutions that leave all parties in a 'win-win' position. The ability to develop such a consensus is contingent on a number of factors, including the availability of time and resources, the willingness of the participants to engage in the process in good faith, a fairly clear set of goals within a flexible framework that allows for the achievement of a consensus, and, perhaps more importantly, a skilled facilitator. This chapter explores the nature of multipartite consensus-building and how this approach was applied in the Whitehorse Mining Initiative process.

Consensus-Building and the Facilitator

Successful consensus-making forums require a neutral third party to mediate and facilitate the dialogue. Given the diverse viewpoints around the table, specific skills and knowledge bases are needed to ensure that participants interact productively. David D. Chrislip states: 'Special skills are needed to lead problem-solving teams and few people possess such skills naturally.'[3] A facilitator's responsibility is to initiate, lubricate, and oversee discussions, but not to judge. Further into the process, this role may include reducing rigidities in the bargaining positions of participants, reconceptualizing issues, offering inducements, and overseeing compliance. A crucial distinction between a mediation process and arbitration or legal adjudication is that the 'product of mediation is not a verdict, but consensus among the actors involved, sensitive to the central concerns of these parties ... mediation can also stimulate discourse and reflection about goals, interests, and values and reciprocal education over the issues at hand.'[4] Mediation is an incipient consensus-building exercise in which success depends fundamentally upon communicative competency.

What bridges must be built if diverse and conflicting interests are aired at the same table? It is clear that negotiation is a process: 'The nature of negotiation is to arrive at the largest mutually satisfactory agreement with any one (and therefore, each) getting at least enough to make it want to keep the agreement.'[5] What is less clear and most important, is that success is to be measured against the problem, rather than against the person across the table defending another value and decision option. Success must be measured by more than signatures on a document at the conclusion of a series of meetings. Mary Parker Follett states that there are three ways to address conflict: domination, compromise, and integration.[6] It is likely that in consensus-building forums, all three approaches are evident. Negotiations can be characterized by division, by creation, or by exchange. At one time, policy negotiations were about winning or losing; this is now changing. In order to realize integrative solutions, two or more objectives are accommodated in a way that recognizes the goals of both interests. In an adversarial society, one that is increasingly litigious, 'differing' is seen as 'fighting': 'Once the battle is joined, ego and winning become much more important than exploring the subject.'[7] Seeking consensus requires distinctive aptitudes and attitudes. For these reasons, one of the first challenges in any consensus-making process is to encourage participants to move away from acting on the powerful urge to vent their own ideas and objectives and move toward focusing on the problem itself. In the case of the WMI, the problem was how to achieve sustainable mining that would serve the long-term interests of the public interest. The role of the diverse participants is to help define that public interest.

In a discussion about participants, it is important to consider the issue of representation. In their book *Environmental Dispute Resolution,* Bacow and Wheeler identify three classes of participants: (1) those who hold formal positions that affect the development controversy, such as public officials (elected and/or appointed) and the developer(s); (2) those affected by the development proposal, such as community groups, Aboriginal peoples, labour, and/or special interest groups such as environmental pressure groups; and (3) a mediator who represents no specific stakeholder or constituency and who participates in the process.[8]

It is the role of the facilitator that is easily overlooked. Facilitators adopt a variety of roles including problem-solving, contending, yielding, inaction, and withdrawal.[9] The importance of their participation in the success of a consensus-based effort should not be underestimated. They are the ones whose task it is to foster communications and the development of new relationships between diverse participants.

Just as the degree of formality in the process will ebb and flow, so, too, do the roles of negotiators. Negotiators must absorb more than the issues on the table. In addition, they must possess an ability to analyze interests, issues, and positions. Many different sets of issues may underlie some interests. The artlike aspects of the process involve allowing 'contradictory views to exist in parallel and then to design a way forward.'[10] Dan Johnston, who served as facilitator for the WMI process and for many other multi-stakeholder forums in Canada, offered the following observations about effective facilitation: 'There has to be a balance between running these processes in a way that is efficient but never compromises the integrity of the process in the eyes of the participants. Facilitators must act as gate-keepers and come up with a product that has some legitimacy with all the participating constituencies. All the perspectives must be represented.'[11]

It is difficult to manage a process that attempts to find some common ground between people with widely diverging world views. During the negotiations, when participants may feel threatened or alienated by the process, the facilitator needs to be able to recognize this and seek to build bridges between interests and ideas. Participants may be redirected by their own constituents and may then slip back to former positions. At that time, the facilitator must attempt to nudge these participants out of their rigid postures. There is, therefore, a need for continued reaffirmation that progress is being made, even if it takes time. Participants often have to learn that it is acceptable to compromise their position as long as they do not compromise their values and as long as they behave in a socially responsible manner. A facilitator who understands the process must make sure that he/she never manipulates people in order to agree to an outcome. There are times when an agreement cannot be achieved. At such times, it is important not to push for an agreement at all costs, but to be satisfied that people have tried all alternatives.

Adaptability is an essential characteristic of a good facilitator. 'Going with the flow' is what Johnston refers to as 'broken field running,' which includes intuition and knowing where to draw the boundaries. One must be able to assimilate material quickly and give it back to the group in a cogent way. A 'passive-aggressive' style may best characterize a good mediator; he or she must not dominate the substantive discussion, but rather be able to refocus the group where necessary.

Finally, but perhaps most importantly, good facilitation is not about ensuring that a quiet, polite conversation takes place. There is nothing wrong with emotion. If people repress their true feelings, it is very difficult to make real progress. Nevertheless, when it is necessary to defuse situations, the facilitator needs to play an active role that might include humour or a variety of other tactics. Johnston suggests that successful processes often happen when members of these groups take some responsibility for themselves and for the outcome. If participants take ownership of the process, the ultimate agreement is strengthened.[12] It is not easy to find the kind of skills and knowledge bases, common sense and intuition, and force and ease of personality that constitute the prerequisites for an effective facilitator. Neither a law degree, nor studies in organizational behaviour, industrial relations, or public policy is necessarily sufficient preparation. As Greg Pyrcz, a professor of political philosophy, observes, the role of the facilitator is akin to that of a 'Rousseauian Parliamentarian,' who constantly seeks to identify and serve the public interest.

The twentieth century has been characterized by a search for certainty and absolutes, a search that is often characterized by heavy dependency on technology and science. A problem with this approach is that the search for alternative avenues is short changed. It is the facilitator's tough role to push for the elusive transformation away from adversarial habits and toward a collaborative mindset. The facilitator must consider the following questions, which are crucial in moving toward consensus:

- What follows?
- What does this lead to?
- What does this open up?
- What are the possibilities?
- Where do we go from here?

The following list includes some of the 'basic critical ingredients' for processes such as the WMI:

- the process is genuinely needed: participants have a desire to meet, talk, and try to work together in a way that is different from previous exchanges
- there is an agreed form for the outcome (policy recommendations, legislative proposal, suggested regulatory procedure, model code, etc.)
- either tacitly or overtly, the process has the support of key decision-makers who will ultimately receive any recommendations that might result
- the right people are at the table and the table is balanced; particular stakeholder groups are not over- or under-represented
- the process is facilitated and managed by an agreed upon neutral person who serves at the pleasure of the group
- there is an agreed upon deadline.[13]

Setting the Stage: Identifying the Players and Establishing the Game Rules

The Whitehorse Mining Initiative did not follow some of the basic rules that are commonly understood to be fundamental to the process of achieving consensus. These include the need to establish clear objectives, carefully designed procedures, and general agreement on the goals and format of the process – in other words, well-defined game rules. On the contrary, the WMI was an open-ended, flexible process that evolved as the initiative progressed. This flexibility proved to be its greatest strength; participants were able to become genuinely involved, help shape the process, and, in so doing, ensure that their diverse interests were accommodated from the beginning. The lack of formal procedures and rules, however, did generate some problems and frustration for the participants. That a final accord was reached said much about the commitment and abilities of those individuals who envisioned a better future for mining and Canada. The WMI process itself reflected an aspiration to achieve a setting that was not characterized by mistrust, entrenched positions, and decision-making paralysis.

It is often maintained that a common flaw in consultative efforts is that those who contribute to the design of the process fail, or neglect, to establish clear guidelines for the process. Such guidelines would include a recognition that participants must agree on the goals, objectives, and format of the consensus-making process and, most important, on the nature of the recommendations submitted at the conclusion of the process. Clarity of purpose allows participants to develop both procedural and substantive strategies. The relationship between the outcome of the WMI consensus-making process and the governmental decision-making process is crucial in moving beyond vision statements to public policy and processes and procedures, where questions of implementation are paramount. In the case of the WMI, getting to a point where an output could be produced required radical changes in participants' trust and understanding. Neither the process nor the responsibilities of the participants were well articulated or defined. As Dan Johnston, one of the main facilitators, observed: 'The process was not conceived up front. It just evolved. There was no teaching in the process about how to achieve consensus. It was a potential time bomb ... For these processes to work well, it is important to plan them well.'[14] The personal commitment and abilities of the players – and their contributions to and ownership of the process – together with the existence of able and influential champions, was the foundation that proved strong enough to support a consensus-based accord two years after the process was initiated.

The idea of the WMI was first formally introduced at the 49th Annual Mines Ministers' Conference in Whitehorse, Yukon, on 22 September 1992. The WMI arose out of a realization that something had to be done to address the problems faced by the mining industry in Canada (see Chapter 2). In the loosest terms, the WMI was based on the perceived need to go beyond government-industry discussions. There was a desire for 'a Canadian mineral policy.'[15] Although the proposal may have been a surprise for government officials, industry's worries were not new to them. As in the past several annual meetings, the Mining Association of Canada (MAC) expressed concern about the uncertain environment within which the mining industry operated in Canada. This time, however, MAC presented a one-page written proposal to convene a participatory consensus-building initiative, involving stakeholders with an interest in the operations and health of the mining industry in Canada. The president of the Mining Association of Canada, George Miller, observed that 'the proposal was loosely modeled on the Canada Forest Accord, which had been signed earlier in 1992 ... MAC suggested highly focused consultations among all the influential policy groups, namely labour organizations, Aboriginal organizations, environmental organizations, industry, and both levels of government.'[16] Once this proposal was agreed upon, the original planning group considered who should be represented and how that representation should be achieved. It was obvious that in addition to the various stakeholder groups, government, as well as industry, support was needed. Furthermore, given its diverse constituency, the WMI was going to require expert facilitation, though this was not recognized until much later in the process.

The Mining Association of Canada asked provincial and federal ministers to support the initiative. In this process, the federal government went along willingly. Natural Resources Canada was supportive (financially and otherwise) and actively involved from the outset, but as a 'partner,' not as a leader, as would have been the case in an earlier era. Government's traditional role in the driver's seat had changed. During the WMI process, a number of participants would crowd into the front seat to take their turn helping to navigate and steer the WMI toward a successful agreement.

Support from a number of key stakeholders within each sector was very important. Walter Segsworth, president of Westmin Resources, observes that the participation of senior industry people and government brought money to the table and then other people were able to sign on.'[17] Moreover, the provincial ministers wanted to support the mining community and such a consensus-making approach seemed like a reasonable exercise that promised a good return as a public relations effort. This was particularly important and timely given the escalating demands for public involvement in decision-making surrounding a decade of constitutional talks among political elites. No one from government left the meeting saying 'no' to the initiative. Three provinces were to serve as a steering committee – the ministers from Manitoba (as a volunteer), the Yukon Territory (as the 1992 host of the Whitehorse mines ministers' meeting), and New Brunswick (as the host for the 1993 annual meeting of mines ministers in Fredericton).

No details were offered at the Whitehorse meeting about the mandate, structure, or time-frame for the consensus-based process. There were many questions to be answered, including:

- Would the initiative take place?
- What would be the over-arching vision and objectives?
- How would the issues be organized or clustered?
- What should be the role, size, and composition of the steering committee?
- Would task forces be necessary? If so, how many would be needed? What would be their mandate and the time-frame for their work?
- Should there be a secretariat to support the effort? If so, what would be its role, size, and location?
- What would be the budget of the exercise, and how would the fiscal costs of the process be shared among the stakeholders?

At the initial meeting, provincial ministers did not push for clarification of the parameters of the proposed initiative, although they did leave the meeting willing to embrace the process. Since the one-page proposal was somewhat ambiguous, the provinces were able to bring to it different perceptions about their roles, responsibilities, and the purpose of the initiative. This lack of clarity about the objectives and procedures – or from a more positive perspective, its flexible and evolutionary character – characterized the entire process right up to the submission of the Whitehorse Mining Accord at the annual meeting of mining ministers on 13 September 1994 in Victoria, British Columbia.

There was some concern about the role of the politicians in the WMI;

more specifically, there were questions about where they would fit if they were involved and the degree to which they would drive the process. It was seen as important that politicians not steer the WMI, because this would undermine the goals of the initiative. Given that ministers are typically involved in bilateral negotiations in which government takes the lead, some participants commented in hindsight that it was difficult at times to get ministers to accept that they were not driving the process. The WMI was to be a process in which the politicians would accept their place among other autonomous groups.

At the outset of the process, one minister asked industry representatives to sit down with officials and identify the issues; such a direction would have moved the WMI away from one in which the diversity of stakeholders would all share in identifying the issues. Some public officials who were interviewed perceived that the process and the product were generally viewed as being owned by governments. At the time, it was unclear whether public officials should participate only as one group among many in a multi-stakeholder process, or whether they were to show leadership as government actors. It is not surprising, however, that several mines ministers quickly and enthusiastically agreed to bring together key stakeholders to identify and address the labour, land access, environmental, and other issues facing the mining industry in Canada. The main role of mines ministers is to serve the mining community. If MAC said that the WMI was needed to address the issues threatening its survival, then the ministers were obliged to ease the industry leaders' concerns by supporting the process. The Alberta government contributed to the first portion of the budget. It did not, however, contribute further, either in fiscal resources or as a participant. The government did not support the initiative. This was fairly easily rationalized by pointing out that metal mining was quite small in the province and any mineral-related activity was undertaken by junior companies. Another possible reason could be that the goals of the WMI did not mesh well with the ideological orientation of the fiscally conservative provincial government, or perhaps the WMI itself did not have the necessary support among key stakeholders.

Another issue that required serious consideration in the formative stages of the WMI was the role of the provinces. How should they be included? Was the WMI to be a national group and/or a number of provincial groups? Once the decision was made that the WMI would be a national initiative, the question arose about how to bring the provinces into the process. Consideration then had to be given to interest group representation. Often, key people in organizations were contacted; they then decided how they were going to be represented.

Quite early in the process there was strong and widespread commitment to the initiative. Concern for the future of the mining sector provided incentive to search out new means to achieve agreement among stakeholders. A draft commitment statement from the mines ministers proclaimed: 'We strongly believe that this kind of approach to issue identification, consensus-building, and conflict resolution is a productive way that lasting solutions will be achieved.' There was also pressure to realize some early progress in the process. Elected government officials, ever aware of their relatively brief terms in office before the next election, argued that some issues should be resolved within the first six months. The remaining issues, it was suggested, could be settled in the following eighteen months. George Miller firmly maintained that 'there was nothing that could be resolved in six months.'[18] That unequivocal vision set the stage for a longer-term process in which more meaningful resolutions could be achieved. The role and responsibility of politicians, however, was an important consideration throughout the WMI. Dr. Dixon Thompson, president of the Rawson Academy of Aquatic Science, observed in his written comments to the leadership council in September 1993:

> There is a great deal of individual and collective accountability resting on the WMI. Having started the process it will be difficult to abandon it if problems are encountered ... However, the circumstances are a little different for the Ministers involved. They are generally very busy and used to having others deal with process, substance, and strategy. In addition, there is a political tradition in Canada that suggests that a Minister cannot be held accountable for actions (or inactions) in an office they no longer occupy. Ministers may be able to move on with less accountability than some of the rest of us.[19]

Despite the best efforts to design the WMI as a forum of equals, it is clear that the roles and responsibilities of different players would remain considerably different in ways that would not be significantly altered.

The initial planning committee involved three ADMs (one from each of the co-chairing provinces), MAC, and the federal government (NRCan). Its members met at a workshop in Ottawa in the fall of 1992 to think about the scope, objectives, and process that should define the WMI. Highest on their action list was to assemble a group of fifteen to twenty members. By November 1992 invitations were sent to various stakeholders. As it was an initiative that broke entirely new ground, invitees were chosen somewhat at random, and also on the basis of their name recognition. A request was directed to the Assembly of First Nations (AFN) asking them to send a representative. A reply was received, reminding the WMI planners that there are four national Aboriginal groups in Canada, and that all should be involved in the process. To solicit participation of representatives of 'the environmental sector,' an invitation was issued to the Canadian

Environmental Network (CEN), which links more than 2,000 groups, the Canadian Nature Federation, and the Rawson Academy of Aquatic Science. The CEN responded with the submission of a *Proposal to Facilitate a Communication Plan for Canadian Environmental Network Member Participation in the 'Whitehorse Mining Initiative.'* In that document, the CEN stated:

> Coordinated, effective ENGO participation in the development of the Whitehorse Mining Initiative is critical to the success of the initiative. Open and effective ENGO participation will benefit the industry as well as the federal government, the provinces and the territories by reducing the political uncertainty associated with the needs of environmentalists. By understanding these needs now, and taking first steps to incorporate these needs into future mining activities, both government and industry will be able to more effectively plan mining development in Canada. This will greatly reduce the costs of developing (and redeveloping) environmental regulations and enhance opportunities for foreign investment in Canada's mining industry.[20]

Individuals who had backgrounds in areas related to mining were invited to participate; those initial members formed the nucleus that facilitated the formation of a larger, broader group of stakeholders. Initial contact to a sector's broad network of interest groups, such as the environment sector's CEN or the Assembly of First Nations, enabled the national working group to cast a nationwide call for participation. The CEN selected participants on the basis of ability and background (looking for those people with experience in the mining sector), as well as on the basis of gender and regional balance.

In bringing representatives from the five sectors to the table, it could not be assumed that they would act as direct representatives because of the diversity that existed within each policy community. Understanding, therefore, began with removing some basic stereotypes people had about other groups. There is a tendency to think of different stakeholders as homogeneous entities such as the 'Industry' or the 'Environmentalists.' In the case of Aboriginal representation, there was an additional challenge that persisted throughout the WMI, which contributed to the unwillingness or inability of the AFN to sign the final Whitehorse Mining Initiative Accord in September 1994. It is difficult to achieve the involvement of Aboriginal people through a representative process. As Hans Matthews, president of the Canadian Aboriginal Minerals Association (CAMA), explains, 'In Aboriginal culture, no one person is representative. My role, and CAMA's role, is to facilitate dialogue with everyone.'[21] Matthews identifies an important constraint on consensus-building initiatives such as the WMI that seek to include Aboriginal leaders:

> Another issue arising out of the WMI process is the difference in the decision-making structure of the mining company/government versus the Aboriginal community ... they are opposite. The Chief or community leader is not the key decision-maker, unlike the company or government where the leaders are decision-makers. A Chief is a 'vehicle' by which community messages or decisions are voiced. How can we establish a WMI Leadership or Working Group comprised of Aboriginal leaders, business and government leaders and make decisions when one participant group can't make decisions at the table?
>
> You can quickly see how the community consultative process (input and decisions) can become time-consuming and costly if done where all community members, youth to elder, participate. A case study illustrating the complexity of Aboriginal community consultation might be the 'Charlottetown Accord' process. The AFN and other Aboriginal political groups spent millions of dollars and many months. Whose messages were the Aboriginal leaders purveying? The Charlottetown process may also illustrate how 'national,' especially politically driven, initiatives are out of touch with the 'grassroots community level.'[22]

Matthews emphasizes that the ultimate success of the WMI will depend on whether its philosophy can be effectively adopted by communities where exploration and mining take place.

Furthermore, as Keith Conn, from the Assembly of First Nations, emphasized, 'We're *not* "stakeholders." That term has been used and abused ...'[23] When the issues to be discussed are education and training or Aboriginal treaty rights, then, according to Conn and other Aboriginal representatives, they are 'nation-to-nation concepts.' From this perspective, what the WMI represents is an accommodation of third-party interests; land access, land claims, and other issues should be dealt with on a government-to-government basis – Aboriginal peoples are not 'sectors' or 'stakeholders' as other interests around the table are defined. Hence, from the outset, the significance of the classification of roles, responsibilities, and relationships through which First Nations would or should be interacting in the WMI was a profound issue for some of their 'representatives.' Other perceptions identified by an Aboriginal representative were that Aboriginal peoples were being invited to participate in the WMI 'out of sympathetic reasons' rather than as true partners, and that could 'throw cold water on the whole process of consultation.' Despite these serious misgivings, Aboriginal peoples were represented at the table. In the words of one representative who was interviewed, 'if not for someone paying the way, we wouldn't have come.'

Another question of representation had to do with the degree to which bodies such as the Assembly of First Nations had a mandate to represent Aboriginal peoples. For example, the difference between an Inuit and a

Dene is as great as 'the difference between Chinese and East Indian peoples. Their traditions and attitudes can be like night and day ... The span within Aboriginal groups is as large as the span that exists between Aboriginals and non-Aboriginals.'[24]

The issue of representation also affects industry. As the process continued, the Mining Association of Canada (MAC) increasingly assumed – by default, almost – the status of an umbrella organization for 'mining in Canada.' This was an inaccurate portrayal of the official role of MAC. MAC represents a particular sector of the mining industry. The association is primarily representative of the large metal-mining companies such as INCO, Noranda, and Falconbridge. Furthermore, MAC does not have significant membership from Canada's junior companies, which are deeply involved in exploration.

Junior companies were represented through the Prospectors and Developers Association of Canada (PDAC). Certainly, their membership in the PDAC gave them a voice in the WMI, especially with the active participation of the managing director of the association, Tony Andrews. Nevertheless, some direct participation by members of the PDAC would also have been productive. Aboriginal issues, community and labour issues, and environmental issues are all policy fields that their work touches. If junior miners had been more widely represented, they would have benefitted from the opportunity to learn about the perspectives of the other stakeholders, to grasp the significance of the initiative, and perhaps ultimately to adopt at least some of the basic principles of the final accord.

Labour was under-represented in the process according to some participants from the other policy communities. Initial contact was made with the United Steelworkers' Union. Bert Pereboom, currently a doctoral student at Harvard University, was one of the representatives selected by the union because it was believed that he would be able to represent labour's views without being easily intimidated. Labour saw a clear niche in discussions about workplace and mining community issues. It did not see its participation from the perspective of larger issues that the WMI addressed. Unlike the environmental sector, labour did not present a spectrum of ideological views. Another participant from the United Steelworkers' Union, Richard Boyce, noted that the structure of the union organization made labour representation easier than was the case for environmental or Aboriginal groups, because the latter groups were not represented through formal structures.[25] Unlike labour, there was a need for the environmental groups to keep checking with each other to ensure that they were clearly representing the various perspectives. The environmental sector consisted of a broad spectrum of views, ranging from those who recognized that there was a place for sustainable development to deep ecologists who did not support development.

There was much to be accomplished in a relatively short period of time. The deadline of 13 September 1994 was omnipresent in the minds of the participants. Furthermore, there was a cultural difference in understanding the meaning of time. For Aboriginal peoples, for example, responsibility is not a short-term consideration, but a generational one; this is indicative of their long-term planning orientation that sharply contrasts with the mainstream North American orientation to the immediate or short-term. With a final product due at the annual meetings of the mines ministers in September 1994, the process reflected a rush to resolution, particularly for some Aboriginal representatives. Keith Conn of the AFN stated that within a two-year period, the WMI was 'trying to change one hundred years of doing business.'[26]

Participants came to the table with some trepidation and wariness about the success of the project. Given such doubt and some of the concerns outlined above, one might wonder why the different sectors became involved in the WMI. George Miller identifies four compelling reasons: (1) an intellectual curiosity; (2) recognition of the need for new processes to resolve issues; (3) the educative promise of the process; and (4) the search for behavioural changes.[27] Alan Young, an environmental representative, states that 'environmentalists did see it as an opportunity to learn how the industry thinks and functions.'[28] In addition, the sectors offered formal reasons for their participation. Table 4.1 summarizes the statements submitted by the stakeholders about their 1993 perspectives on the WMI and what they hoped it would accomplish for them.

The National Working Group

At a first meeting (later dubbed the 'national working group' meeting) during 10-12 February 1993, there was fairly strong regional representation from across Canada. Participants included representatives from the following associations and governments: the Canadian Nature Federation; Energy, Mines and Resources Canada (later renamed Natural Resources Canada); the governments of Manitoba and New Brunswick; the Inuit Tapisarat of Canada; the Metis National Council; the Mining Association of Canada; the Prospectors and Developers Association of Canada; the Rawson Academy of Aquatic Science; and the United Steelworkers of America. Absent were the representatives from the Assembly of First Nations and the Native Council of Canada.

It was generally agreed that the outcomes of the WMI should include:

- a strategic vision for the minerals and metals sector
- a series of 'accords' or 'partnership agreements' among various stakeholders and industry
- options and recommendations for policy or regulatory changes, consistent with priorities in this jurisdiction

Table 4.1

Stakeholder statements at the WMI

Aboriginal groups see the WMI as an opportunity to 'seek widespread consultation, share information – particularly about the aspirations of Aboriginal peoples – and present key issues with all stakeholders.'

Labour representatives indicate that 'citizenship is what ultimately underpins labour's commitment to the process embodied in the WMI. Labour views its roles as "stakeholder" in broad social terms, although they are terms rooted in our primary role as representatives and advocates for the driller, driver, word-processor, and labour-technician ... sharing the economic benefits means sharing the responsibilities.'

Environmental organizations believe that 'The situation of decreasing investment in mining, unresolved land claims between Aboriginal and federal and provincial governments, environmental degradation, and a vulnerable labour force, requires a comprehensive and coordinated response by Canadians. New partnerships must be forged to ensure international and domestic investment in an ecologically, socially, and economically sustainable mining sector.'

Industry bases its expectations on overcoming the current reality that 'the sector is poorly understood, is pursued by historical images, and faces a rapidly changing investment, environmental, and regulatory climate.' While the 'WMI will not solve every problem nor end future controversy, it can open avenues of dialogue, increase understanding and appreciation, and facilitate behaviour changes on all sides.'

Governments, in recognizing that every province and territory of the country benefits from mining, want to join with other stakeholders to assist the mining industry to fulfil its dual responsibility – to be a strong and growing economic contributor and to be a steward for the natural-resource wealth we all enjoy.

Source: Whitehorse Mining Initiative Secretariat, Workbook for Participants in the Whitehorse Mining Initiative Leadership Council Luncheon Meeting, 14 September 1993, Section 4 – WMI Vision Statement, 2-3.

- establishment of new consultative mechanisms to ensure the continuation of communication and cooperation, and to facilitate the management and/or resolution of conflicts.

At the meeting, a time-frame was established for the process; it was not to take longer than twelve to fifteen months, concluding no later than the annual meeting of mines ministers in the fall of 1994, with the hope that some announcements of progress would be made earlier. Participants recognized the need for selective research, discussion and position papers, interim communication processes, and a number of limited conferences and workshops to facilitate consultation.

The organizers recognized that success would be contingent on the fulfilment of a variety of goals: (1) to focus on root problems rather than merely on the symptoms of issues affecting the mining industry and its stakeholders; (2) to identify and respect the different views, values, experiences, and perceptions that would be represented around the table; (3) to respect the provincial, territorial, and regional level activities and approaches; (4) to contain the costs and complexity that might cripple the process by focusing on the finite limits and goals for the WMI; and (5) to maintain the commitment to involve a broad base of stakeholders.

Appropriate representation is particularly challenging when there is no umbrella group that has a mandate to act as a representative for a constituency. Accountability relationships are similarly difficult, as much for governments as for groups such as Aboriginal peoples. The lack of involvement of Alberta public officials, for example, may reveal those public officials' sense of accountability to their junior companies, who tend to be suspicious of government intervention. In general, prospectors and developers would prefer incremental change by a single government actor rather than leaving themselves open to some agreement by four other sectors setting out new abstract relations and objectives. During the WMI, accountability relations were more complex in some jurisdictions than in others. For example, responsibility for policy issues in the territories may rest with the federal and/or the territorial government depending on the issue area.

The nature of representation and accountability of the various policy communities would also be affected by the financial resources available for funding participation in the consensus-making initiative. In a period of fiscal restraint, funding of the initiative was a serious challenge. 'A shared decision-making process is relatively expensive because it attempts to include all affected interests in a meaningful way. The investment is justified on the basis that shared decisions are likely to be more fair, stable and reflective of the participants' interests.'[29] The national working group went to the 10-12 February 1993 meeting of stakeholders with a budget of $1.3 million, but governments balked at that amount. Participants were sensitive to the need to undertake consultation at the regional level and to support national level activities. It was expected that a minimum budget of approximately $750,000-$800,000 would be required to support such efforts at the national level. Moreover, it was thought that the provinces and territories would pursue and support their own regional and local level processes that were not included in the national-level budget.

The WMI was formally accountable to three funding partners: the federal government, which assumed one-third of the costs; the ten provincial and two territorial governments, which assumed another third (split according to the ratio of their mineral production); and the Mining Association of Canada – industry – which assumed the remaining third. In this process, the government of Canada played a vital role. It did not lead in the traditional sense, but it did provide the 'glue' to keep the process together. It performed a lot of backroom work that would not otherwise have been done. In an era where the Canadian public is increasingly rejecting the decision-making style and processes associated with executive federalism, the WMI served to redefine the federal role. It remained an integral player, but it acted as much more of a partner than has previously been the case.

The provincial governments were sceptical about the process, particularly when their fiscal contributions were established. A major turning point to their concerns was highlighted by a deputy minister who stated that at issue was 'not whether we can afford to participate, but whether we can afford not to.' All the provinces contributed to this first fiscal tally. There was a clear recognition that the consultative process could only proceed if financial barriers to participation were lessened to create a fairer playing field between the sectors.

Several Aboriginal representatives observed that their associations did not possess the technical and other skills and resources needed to participate effectively. Keith Conn of the AFN stated:

> I did not have the technical and scientific staff in house to do proper analysis and research to formulate sound policy and recommendations ... Government and industry assumed we (First Nations) have a bureaucracy we could access automatically, as they do ... I did not have the resources to have access to a First Nations forum/body for technical and political direction. The resources were not available to do preparatory work, research, analysis of options, examination of existing case studies and finally, development of recommendations.[30]

Another participant, Hans Matthews, explains that even if the CAMA or another Aboriginal association had the resources to liaise, share information, and receive responses from their constituent communities, 'those communities' responses may not be representative of the community as a whole and their direction and interest in the minerals industry.'[31] The time and other resources of Aboriginal communities are largely spent reacting to health and other survival issues, with very few resources directed toward education and economic development. These information and knowledge gaps, and the shortage of human and financial resources, pose enormous obstacles to participation. In this context, how can it be possible to report to and receive [informed] direction from one's 'constituents'? Financial contributions, therefore, to support participation in a process like the WMI, are an important resource if the goal is to encourage equal participation from different groups. This is also the case with other sectors, such as the environmental communities. Alan Young states: 'One of the advantages of the process was that it properly funded the environmental participants to take the necessary time for review and discussions with other activities.'[32] Among the stakeholders interviewed, there is a general consensus that the provision of funds is required to allow some constituents to participate. Another possibility was also expressed, however, that some groups, such as labour, could have acted as equal stakeholders by paying their own way if they could. However, as a general principle of consensus-based processes, it is crucial to ensure that stakeholders have the resources necessary to participate effectively, and hence, it may be necessary to support their involvement.

In the case of the WMI, the very basic resources for inclusion were necessary, namely, the funds required to attend the sessions, let alone other resources such as staffing and research tools to prepare for the meetings. In looking at questions both of affordability and fiscal accountability, it should be noted that the budget does not reflect the travel costs of government and industry representatives, including airfare, which were assumed by industry and government. Consensus is costly.[33] Each of the voluntary sectors was allocated a budget. It was clear that there was going to be a shortfall in the secretariat's budget. A total of $454,000 was provided to support participation by stakeholder groups.[34] A strategic decision was made to show some results first, use the budget, and when/if it ran out sometime in December 1993 as expected, then the shortfall would be shown and further funding sought.

The Operating Structure

In addition to adequate funding, a complicated national multi-stakeholder initiative such as the WMI requires a well-structured, adequately staffed and funded body at the centre of its nervous system to direct information and communication flows. Early on, issues of design and process were not yet considered in any substantial manner. The MAC and its community entered into the WMI with little understanding of what the exercise would entail, both in terms of process and the substantive issues that other stakeholders were bringing to the table. For that matter, industry representatives were not initially aware that this process would be fundamentally distinct from bilateral negotiations or labour negotiations. They also appeared to underestimate the ability of stakeholders to identify and to articulate their own interests clearly and persuasively. Furthermore, many mining representatives did not recognize the extent of the very different values others held with regard to natural resources. For example, one observer stated that the 'industry was naive about the depth of the environmental community's concerns and they ended up in a much tougher and complex process than they expected.'[35]

Some advances, however, were made toward gaining a better idea of the kind of process that was needed. In particular, the following objectives and considerations were identified:

- a need to ensure high-level profile and commitment from all relevant sectors and major groups across Canada, by demonstrated leadership at the most senior levels
- a need for a small, effective but broadly representative body at the working level, to provide overall planning and coordination for WMI
- a need for a small number of specialized groups ... to focus attention and marshal specialized knowledge and resources on selected specified issues
- a commitment to ensuring maximum possible consultation and input (within available budget and time), at both the national and regional/local levels
- a commitment to utilizing existing resources to the maximum extent possible, so as to facilitate timely input and minimize duplication and overlap
- a recognition that particular issues and related needs, priorities and circumstances, will vary appreciably from one region and jurisdiction to another, and will therefore require flexible approaches at the regional and local levels.

Progress had been made to cluster the issues into five broad themes, listed in Table 4.2. Aboriginal representatives argued against a separate 'Aboriginal group,' maintaining that they had concerns in each of the other four policy clusters and that 'they had been studied to death.' It was agreed not to have a separate cluster field of Aboriginal issues and, instead, that each group would identify and address policy issues relevant to Aboriginal peoples. The clusters of issues also reflect the fact that the WMI was an issue-oriented rather than a goal-driven process. The WMI was an evolutionary design process that came at the expense of strategic planning. Yet, there are merits to a flexible design process, because it is adaptive to the needs of a unique group of participants who have never worked together. Doug Hyde makes the following astute observation: 'What will pull people to the table? Not goals, but problems and issues! ... The process emerged in a very wise way.'[36]

The Whitehorse Mining Initiative did eventually adopt a working operating structure (see Figure 4.1). A leadership council was formed to spearhead the initiative. The work of the council was supported and coordinated by the WMI working group. Four issue groups were formed to address four

Table 4.2

Thematic issues for discussion at the WMI

Land access, land use, land allocation	Aboriginal issues
protected spaces	participation
biodiversity	training
jurisdiction	tenure
tenure	self-governance
ensuring access	jobs
Aboriginal land claims	operating agreements
resource evaluations	cultural awareness
concept of economic rent	self-determination
permitting	
	Financial performance/taxation issues
Workforce/workplace/community issues	income-tax issues (exploration incentives, taxation of reclamation funds)
training	mining tax
safety/health	non-profit taxes
standardization of regulations	mandated costs
productivity	capital investment
pensions	international trade
education	international standards
adjustment	cumulative effects of taxation on mining industry
community stability	
job security	
employment equity	
Environment issues	
assessment process	
regulatory regime and process	
self-regulation	
reclamation	
environmental commitments (on a global scale, including international agreements)	
regulatory process	
permitting	
standard setting	
codes of practice	
technology	
liability	
harmonization of regulations	

main areas of interest that were considered important to the industry. Each of these bodies was to have representation from all of the major stakeholder groups. Finally, the secretariat would play an overall support and coordinating role and assist the issue groups in preparing their final reports.

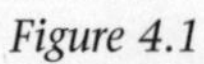

Figure 4.1

Conceptual diagram of the WMI

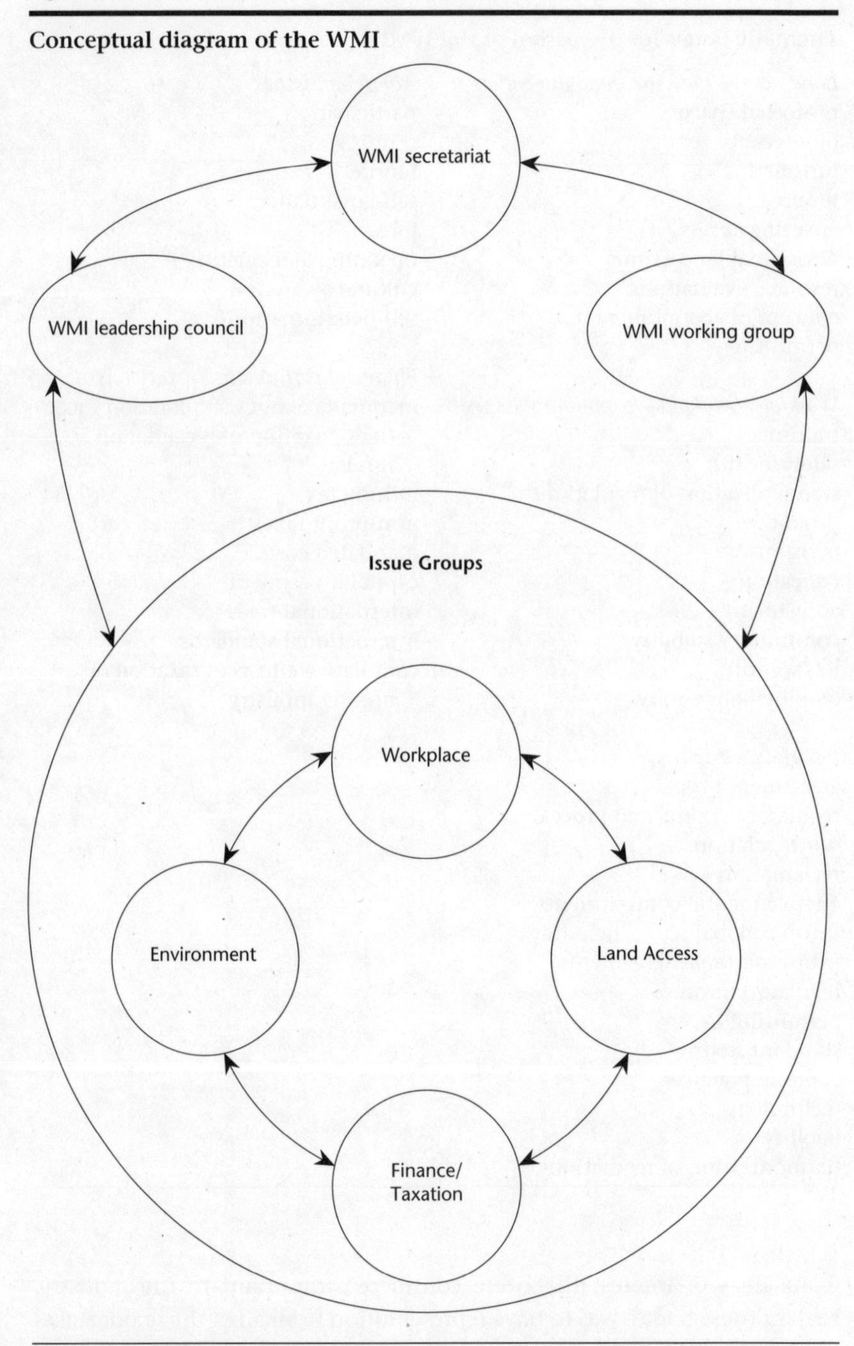

Source: 'Whitehorse Mining Initiative in Full Swing: Working to Promote Greater Understanding,' *PDAC Digest*, Winter 1994.

While the WMI organizers recognized that a permanent secretariat was needed to coordinate the initiative, its role was not clearly defined. One participant stated that the job of directing the secretariat was extremely difficult and that Lois Hooge, who was engaged to direct the secretariat, deserved credit for pulling things together. There was little clarification about what the secretariat's role was to be in establishing and maintaining communication flows between the communities of interest across the country. Was the secretariat responsible for maintaining communication flows between the thirty-seven people of the leadership council and the approximately seventeen members of the national working group? Was it also required to keep the mining ministers informed? In the case of the provinces, they had all contributed to the secretariat's budget and, therefore, maintained that the secretariat's responsibilities should be extended to include the dissemination of information. Manitoba and New Brunswick, as co-chairs, could have taken up the responsibility for informing the other provinces, but this did not happen. As far as the WMI was concerned, the process continued to be plagued by the magnitude of the task, as well as the diversity of perspectives on issues and problems that were represented around the table. The process was very different from the bilateral meetings with which mining industry representatives and mining ministers were familiar.

Ultimately, it was decided that the leadership group was to have the following composition:

- federal/provincial/territorial mines ministers (13)
- federal minister of Indian and Northern Affairs (1)
- labour (4-8)
- environmental organizations (4-8)
- industry (8)
- Aboriginal organizations (8)
- other leaders and appropriate organizations
  Total: 38-50[37]

It was left to the secretariat to identify other leaders whose candidacy would be considered by the national working group. The leadership group was to be present, nominated, and announced at the March 1993 conference of the Prospectors and Developers Association of Canada (PDAC). At that meeting a leadership council was established, and it reflected the diversity of the stakeholders. The forty-member leadership council consisted of mining ministers, industry leaders, and representatives from environmental interest groups, labour, and First Nations. The leadership council 'was originally intended to provide a forum for accountability and reporting, give advice, demonstrate commitment, safeguard the integrity of the process, and assist in resolving substantive issues while promoting implementation of the results.'[38] George Miller explains the strategy: 'The structure of the WMI was not a conventional hierarchy. Different groups served different functions, but one group did not 'report' to any other group. Each structural unit had multi-stakeholder membership and a separate function, but none of the bodies had authority over the others.'[39] The process evolved under a full array of expectations. Up to this point, most participants still did not recognize the magnitude of the undertaking.

The sixteen to twenty members of the working group were stakeholders' representatives at the management level. They 'planned and managed the WMI process, prepared the expenditure budget, raised funds, assisted in resolving process issues, and directed the work of the Secretariat.'[40] The group met approximately once per month. The first six months, from September 1992 until March 1993, was rather slow in terms of activity. Without precedent, the designers of the WMI, in their fledgling working group, worked hard to get five different sectors to 'buy in' to the WMI, to finance the process, and to set things in motion. It was a long first stage. By the summer of 1993, there was still little understanding of what the secretariat's roles and responsibilities should be and of the boundaries of the initiative itself. Despite this strategic gap, work began during the period of April-June 1993: the working group planned, identified some issues, set a broad mandate for the issue groups, and established the budget. At this early stage, the working group was comprised of several senior people from each of the sectors who met during that period about once every three weeks. Twenty to thirty professionals or experts comprised each of the four issue groups that addressed four major themes of the WMI: environmental issues, financial and taxation issues, land-use issues, and workplace and workforce issues. It was determined that Aboriginal issues should be considered within each of these four issue groups, rather than within a group of its own. Over the two-year period of the WMI, the issue groups met between six and ten times each, for two to three days per meeting. Conference calls, sub-group meetings, and bilateral discussions occurred between meetings.

It was by design, although some people may say that it was by lack of design, that the four issue groups (Finance and Taxation, Land Access, Workplace/Workforce, and Environment) had the autonomy to organize themselves and to determine what process they would use to work toward their responsibilities. They did not report to, and were not subordinate to, the leadership council. The 'symbolic and moral authority' of the representatives on the leadership council was acknowledged, but 'they did not take their orders from them.'[41] Each issue group developed different cultures and group dynamics, which emerged to reflect the policy issues and the personalities, aptitudes, and attitudes of the group's members, and their processes varied as well. The challenges to reaching consensus were enormous. Representatives from five sectors – each with radically different values, attitudes and aptitudes, experiences and perceptions – would have to try to set aside a number of perceptions, including past negative impressions and relations they may have experienced with each other in disputes; stereotypical views they held of each other; and misinformation, as well as information, and knowledge gaps. The distinctiveness of the WMI was that it brought stakeholders, who held such different values around the table, to stake out some common ground and agree to some base level of consensus.

The lack of initial structure in the issue groups was compensated for by the flexibility it gave them to tailor their own structures and processes in a way that would best enable them to identify and broach the issues. The evolutionary approach allowed each group to respond uniquely to policy issues that ranged from highly technical taxation issues to the contribution that Aboriginal people's traditional knowledge bases could offer integrated resource management approaches, as well as to the distinctive group cultures that reflected personalities and knowledge bases of the group's members. Had the working group or leadership council developed and imposed structures and processes, the issue groups may have been too constrained to reach consensus. As one public official reflected, in the early stages of the WMI, the leadership council could have itself identified the issues to be tackled and then asked the issue groups to delve into them. But the process of identifying the issues was a crucial step toward finding common ground. Certainly, the output of the issue groups – the breadth and the depth of their work, and the surprising degree of consensus – indicates that the working group's approach of letting the issue groups define their own issues was effective.

The context created by the WMI is unique since, typically, we tend to surround ourselves with like-minded people who share our values and orientations. During an interview, the minister of Natural Resources in Nova Scotia, Don Downe, suggested that we tend to be overconfident in the knowledge base we think we possess, and we tend to underestimate seriously how much we do not, in fact, know. Over time, there is a tendency to seek out other individuals and select information that confirm our own view of the world and the problems we face. The value of the WMI, according to the Nova Scotia mines minister, was that through the participation of a diverse group of stakeholders, participants considerably broadened their own knowledge base.[42]

Entering such an unknown arena will cause uncertainty, and most of those people interviewed observed that they entered the new playing field with some scepticism and some nervousness. One participant observed that 'people rarely value diverse points of view. Not many enjoy having their view of the world challenged by dissenting opinion.'[43] It is much more comforting and much easier to hear complementary or confirming opinions. It takes much more energy – and emotion – to clarify, substantiate, and re-assess one's own assumptions, information, and knowledge bases. It is, therefore, not surprising that most of us work hard to avoid such encounters, particularly when they are set within adversarial rather than collaborative contexts. The task set out for the participants was to reach common ground on a plethora of issues they themselves would identify. Reflecting on the process that led to the delivery of the accord to the annual mines ministers' meeting in 1994, Doug Hyde identified the following characteristics that define 'good participatory behaviour.' An effective consensus-building member would be someone who:

- is interested in a positive outcome and understands the wisdom of that outcome
- is open-minded
- is willing to listen
- has a vested interest in the outcome
- knows his or her constituency and how to get it to perform
- is articulate
- understands that he or she has a fairly broad role in the initiative.[44]

One might also add: is willing to do his or her homework. Participants in the WMI spent much time sharing, preparing, and reading background papers and policy reports. Each group was able to structure its own distinctive discursive context and to foster the kind of behaviour and attitudes that would encourage participants to learn from each other and to strive toward a shared vision. As the discussion of the work of the issue groups and the leadership council demonstrates in the next chapter, the ability to shape their own environment helped participants move toward consensus and, ultimately, the final accord.

5
The Whitehorse Mining Accord: The Search for Consensus

The work of the issue groups and their subsequent reports served as the 'paving stones on the path to Leadership Council consensus' and the final signing of an accord[1]. The issue groups wrestled with many different issues, competing value structures, and individual biases. It was within these meetings that the process of conciliation and consensus was initiated. It was here that the substantive issues related to the mineral industry were addressed. Each issue group produced reports that included a set of principles and objectives and a list of over 150 recommendations. The reports themselves address four categories of issues to be considered: workplace, environment, land access, and finance/taxation. A draft accord was then drawn up by a new group based on the reports of the four issue groups. The final accord contained a vision statement, sixteen principles, seventy goals, and a statement of commitment ensuring that action would be taken toward implementing both the vision and the goals.

The Finance and Taxation Issue Group

The Issues and Recommendations

The Finance and Taxation Issue Group was first off the mark, with its initial meeting in Toronto on 30 July 1993. The work of the group was roughly categorized under the broad areas of capital formation; the costs of doing business (taxes, charges, and regulatory compliance costs); mine reclamation; government services; and aboriginal-industry business relations.[2] Many of these issues have been discussed in previous chapters (see, for example, Chapter 2).

In the area of capital formation, the issue group focused on the importance of equity capital as the only source of capital for junior companies; exploration companies are considered vital to the discovery of new economic deposits in Canada. Investment capital is also a requirement for infrastructure development essential to the development of mines in remote regions. Yet the ability to attract investment capital is threatened by a variety of uncertainties, including concerns about environmental liability and erosion of mineral title. Recommendations were made to reduce uncertainties caused by the regulatory environment, to settle the issue of Aboriginal title fairly, equitably, and as quickly as possible, to improve security of mineral title, and to generally provide an environment that will improve industry's ability to raise equity and debt capital. In terms of exploration investment capital, one of the recommendations was for the federal government to introduce a national exploration incentive program to allow Canada to remain competitive and to counteract the decline in base-metal reserves.

It was argued that the cumulative costs of doing business (mining taxes, charges, and regulatory compliance costs) were becoming a 'prohibitive' burden on mining operations. Moreover, it could be difficult to open new mines if the associated costs do not render an attractive enough return on business. It was suggested that the tax regime needed to be changed in a way that was perceived to be 'simple, pragmatic and fair.' A simplified mineral taxation system, with harmonized federal and provincial tax policies, would help the industry attract investment.

Mine reclamation also presents a significant additional cost to the industry. Nevertheless, mine reclamation was unanimously recognized by the committee as essential to protect the environment and public health and welfare. The committee suggested that steps could be taken to ensure that adequate funds are available for full reclamation and that the industry's cost structures are efficient. It was recommended that flexibility in the financing of reclamation be provided through a variety of funding options to help companies meet the necessary requirements. In the area of abandoned mine sites, efforts should be made to establish responsible parties. There is also a need to make sure that a fund is in place (generated by revenues and taxes) that can pay for clean-up of abandoned sites. Furthermore, it was recommended that an individual or a company that is exploring an old mine site should not be held responsible for environmental damage that it did not cause. Persuasion could also be applied to industry to encourage them to clean up orphaned mine sites voluntarily.

Government services and support of the development of mineral resources is an extremely important variable in determining the ultimate success of the mining sector. In an era of declining budgets, industry is concerned that governments' ability to provide essential services will be undermined. It was recommended that government attempts at cost recovery and revenue-earning activities should not jeopardize the provision of those essential services.

Aboriginal-industry business relationships have been adversely affected by poor communications and misunderstandings on both sides, uncertainty regarding the areas affected by land claims, and various impediments (legislative and otherwise) that have inhibited the formation of effective partnership arrangements. The issue group suggested many legislative and policy changes, financial incentives, and improved communications that would encourage the participation of Aboriginal people in mineral development. One of the suggestions was that Aboriginal development organizations could be used more often by Aboriginal peoples to raise their profile in the industry and attract investment. Government could also create some fiscal incentives that would facilitate the formation of joint ventures between Aboriginal people and industry.[3]

The Process of Decision-Making

Members of the Finance and Taxation Issue Group decided to proceed as a core assembly rather than break into sub-groups to tackle the issues. That approach would allow them to deal more effectively with the issues, to manage the flow of information from invited experts, and to ensure participation from a diversity of stakeholders. Members agreed to the following set of guidelines for their deliberations – their rules of the game:

- All have the responsibility to listen carefully and attempt to hear each others' points of view, and to treat each other with respect.
- All decisions will be made by consensus.
- Dissenting views will be recorded in the minutes. It will be the responsibility of those with dissenting views to present them clearly, to describe the area of disagreement, and to propose alternatives.
- Unsettled issues may be dealt with by further dissertation and investigation, referred to another person, or left unresolved at the discretion of the group.
- Every member will be expected to come to each meeting adequately prepared to discuss the issues to be considered, to flag any problems, and when appropriate, prepare recommendations.
- The exact date and location of each meeting will be determined at the prior meeting. This way the group will remain flexible and can reflect the requirements of the core members.[4]

The Finance and Taxation Issue Group was run like a business meeting – not surprising given the composition of the group; for example, the Aboriginal representatives were senior executive (vice-presidential level) bankers. It was interesting to hear one participant who was interviewed observe that the Finance and Tax group's culture was best defined as 'polite.' This dynamic is reflected in the group's explicit recognition at the outset that its members could not, and should not, 'commit themselves or their superiors to any course of action recommended. This will encourage a free ranging discussion and expression of views.'[5] One participant stated:

> Our group decided right from the very beginning that each member could only present their personal experience – they couldn't negotiate on behalf of, or represent their constituency – we would never have got anywhere when it came to sensitive areas. When the issue group recommendations floated up to the leadership council, then it was through that process that the representation or constituency part of the process was dealt with. It couldn't be 100 per cent democratic.[6]

These statements reflect a step toward pulling participants out of their organizational role, which should leave them more open to examining issues from a variety of perspectives.

Participants at the Finance group's next meeting on 13 August 1993 included public officials, typically at the director level, from Finance Canada, Department of Indian and Northern Development, Energy, Mines and Resources Canada (now Natural Resources Canada), and the provinces and territories. As well, there were senior industry officials (CEOs) from major mining companies. Those core members reiterated the need they had expressed at their July meeting to have a membership that reflected the spirit of the WMI process. They reviewed their membership and found that 'an environmental representative was of the utmost priority';[7] the CEN had not provided a name so the group sought out a representative from Environment Canada. Other people were also approached to join the group. As well, the group sought to have representatives from Quebec and from the labour (in July 1993 labour chose not to participate in the Finance and Taxation Issue Group), banking, and brokerage sectors.

At the August 1993 meeting, the Finance and Taxation Issue Group addressed several objectives using the following methodology:

> Several members, prior to the meeting, were contacted and asked to write a short synopsis on one of the seven issues identified at the first meeting. The outlines consisted of a short statement defining what the particular issue was; and what would be the group's objectives regarding that particular issue. The members used these initial statements as a starting block. From there the group developed their own statement – clarifying the issue for all the members – and then set down their objectives for that issue.[8]

The Finance group deliberately did not hire a facilitator, choosing Tony Andrews of the Prospectors and Developers Association of Canada to chair their group. This decision was made, at least in part, because of the complicated, convoluted, and technical nature of the issues the group had to tackle. It was felt that an effective chair would have to possess some depth of knowledge about the issues. Several participants who were interviewed found that the group was well managed. It was stated that Andrews's success as chair stemmed from his recognition of his biases and associations. Having a member of a vested interest group chairing such a process is not usually advisable because participants may not feel that all their perspectives will carry equal weight in the discussions. Andrews's knowledge and understanding of the issues, however, helped facilitate the process.

The finance group was also successful because the group's dynamics were defined by the technical nature of the issues; their meetings were very structured and they gathered facts and invited experts to make presentations to the group. It was the group's chair who took care of most of the writing, wrote up the meetings' minutes, and took them back to the group to confirm that the minutes accurately portrayed the discussion.

Perspectives other than those of industry and government were presented at the Finance group's meetings. Whether or not government should establish a national exploration incentive program is an example of an issue that preoccupied the group. Due to the involvement of players other than government and industry, bankers and environmental representatives were able to ask questions that helped broaden awareness of the considerations to be addressed in creating such a program. Their questions about the nature of different programs and their search for alternative ways in which programs might operate were fundamental to the group's achievement of a fuller discussion of the issues. Quite a broad range of concerns were addressed, in contrast to the narrow focus of issues under discussion had the participants been confined to industry and government.

A good example of this was provided by the continuing debate between government and industry about the mining community's perception that Canada is a high-tax environment in which to do business and that the climate discourages international investors. A review of the issues revealed that the perception emerged from the speeches of CEOs of mining companies. Discussions in the group ultimately revealed some different perceptions of the issue; some contended that Canada offers a pretty favourable tax system from an international comparative perspective. Similarly, the group's efforts revealed to industry representatives that the kind of tax incentives that industry requested were simply not possible, because government cannot favour one sector over another. As a result, it was noted that the approach was less confrontational. Although from time to time some members were pursuing their own agendas, it was the diversity of players, and their willingness to listen and learn, that contributed significantly to finding common ground.[9] In an interview, Bill Toms, chief of Resource Taxation, Business Income and the Tax Division in the Tax Policy Branch of Finance Canada, acknowledged that he had approached the first meetings with the concern 'that this was just another attempt to lobby, and that my participation would be a replay of meetings [between industry and government] over the last five years.'[10] Toms indicated that 'while several participants did arrive at the meetings with specific objectives, because others who had no specific objectives were involved, we spent more time on other issues which might have been otherwise neglected.'[11]

The Land Access Issue Group

The Issues and Recommendations

Land-access issues were undeniably the most difficult on the WMI agenda. From the outset, the group was variously characterized as 'quite dynamic' to a group with 'real tensions.' The land access group members prefaced their final report with the caveat that they were not speaking on behalf of their respective organizations but as individuals. The document that they produced represented their consensus views. The report was divided into sections dealing with land access and mining; Aboriginal land claims and interim measures; completion of Canada's protected areas networks; and land-use planning and decision-making processes (see Chapters 2 and 3). The recommendations this group reached are well worth noting. Many of the participants were surprised that they were able to come to such a compromise.

In the area of land access and mining, it was suggested that the industry continue to improve its record of environmentally responsible exploration and mining and perhaps investigate the idea of the 'feasibility' of a model mine program. A recommendation was also made that governments develop and communicate some clear policies about the issuing and/or cancellation of mineral tenure and the associated issue of compensation.

Unsettled Aboriginal land claims and interim measures generate a great deal of uncertainty both to Aboriginal peoples and to the mineral industry. It was recommended that governments concentrate on the 'expeditious and 'efficient' settlement of land claims. Furthermore, the structure of negotiations should be well defined, understood, and clearly communicated. This should also apply to land-access provisions contained in settled Aboriginal land claims agreements. All non-confidential information regarding Aboriginal land claims should be readily accessible to all stakeholders. As recommended by the finance and taxation group, governments should help Aboriginal peoples to pursue interim business agreements with mining companies.

Achieving consensus for the completion of Canada's protected areas networks was a fundamental goal of the environmental community. The issue group endorsed that goal as stated in the 1992 Tri-Council Statement of Commitment to Complete Canada's Networks of Protected Areas. In the determination of those protected areas, it was recommended that all stakeholders be able to 'participate meaningfully' in the process. In the case of mining, it was recommended that Mineral Information Inventories (MII) be conducted and evaluated before selection of the final site in order to ascertain the relative potential for mineral development of different areas. This issue is a subject of debate. MIIs are used to define areas with high mineral value. From such inventories mineral and supply resource assessments can be conducted. Such studies are time-bound, because as new uses and technologies for extracting ore become available and as world prices change, the potential value of an undeveloped deposit will also change. It is, therefore, difficult to make an accurate assessment of the future economic potential of a particular area in order to make long-term planning decisions. If MIIs are to be used in making land-use decisions, there is a risk that the information will be inaccurate and valuable mineral areas may remain undeveloped with the land use designated for some other purpose. On the other hand, if such data is not used, decisions may be made to alienate land from mining, because the MII data is not available to help planners make more informed decisions.

In the context of land-use planning and decision processes, it was suggested that governments move toward integrated resource planning and decision-making. Planning should address both ecological and socioeconomic issues. Improved mineral resource information should be made available through industry and government partnerships so that informed land-use decisions can be made. Again, it was suggested that a complete mineral information inventory be conducted prior to any land-use decision. This is very important to the mineral industry. If mineral activities are to be affected by constraints imposed as a result of protected areas, the nature of those constraints should be made clear early in the decision-making process.[12]

The Process of Decision-Making

Given everyone's expectations that the Land Access Issues Group would involve challenging debates and some heated conflicts, they agreed to hire a facilitator, unlike the Finance and Taxation Issue Group. The concepts that necessarily underpinned the discussions and the personalities involved contributed to a rather fractious environment. The Land Access Issue Group was comprised of people with a heavy planning orientation. The approaches of multi-stakeholder representatives can be radically different from one another; the orientation of land-use planners, with their theoretical emphasis, for example, is very distinctive from the practical dispositions of the prospectors. Planners have a specific way of looking at the world, a 'land-use persona,' culled from their own accreditation and their theoretical orientation. Prospectors brought a completely different orientation to the group. The language, symbols, and professional backgrounds of a diverse group of stakeholders make it particularly challenging to find common ground. A radical environmentalist in the group brought a 'deep ecology' perspective; from that perspective, an environmentally friendly mining industry could be viewed as an oxymoron. One member from Parks Canada also brought a 'biocentric' perspective to the discussions. Such backgrounds make it difficult to establish a discursive context that would foster a shared understanding.

For some participants, the Windy Craggy issue in northwestern British Columbia was the flash point for the Land Access Issue Group. This was an area of a proposed mine site that was deemed to be mineral rich with huge mining potential. Environmentalists and others pointed out that the land in dispute had tremendous ecological value. It was recently designated a World Heritage Site and became a protected area (see Chapter 2). This issue continues to be a sore point with many individuals in the mineral industry. Some participants who were interviewed said that the group's members did not want to talk about it, yet their discussions were underscored by it. Windy Craggy really set the context for the group; for example, some group members, according to one participant, 'knew each other from press clippings.' One participant who talked to environmental and industry representatives separately perceived that their views were not incompatible, but the history of Windy Craggy was seen as too explosive to address. Another participant who was interviewed disagreed, saying that Windy Craggy did not dominate the group: 'All sides agreed this was an example of a bad process ... Environmental groups agreed it was not a good process.'[13] Some group members thought that they should not talk about Windy Craggy because they would yell and scream, but one participant noted that 'people found ways to yell and scream anyway.' He thought that 'if Windy Craggy had been examined from the perspective of looking at the decisions and processes that led to the fiasco – there were ways to talk about it which could have been used to focus the group's efforts ... It was a very rich way to understand how we could achieve a better process. An understanding of what happened there could have been high risk [for the group] but people fighting about it without mentioning it was just as disruptive.'[14]

The diversity that characterized the group also was reflected in their understanding of the needs of the group and the roles of its players. Even when they meet face-to-face, people within a room will view things differently. It should also be noted that, in general, the mining industry is often characterized by individuals who are direct, fairly aggressive, and who speak their minds. They can identify with others who do the same.

The Environment Issue Group

The Issues and Recommendations

The work plan of the Environment Issue Group focused on a mine's life cycle: assessment, operations, and closure. The work of the group was categorized according to the following topics: environmental review processes, operations, closure, public involvement, information use of science, and overlap and duplications. The Environment Issue Group also met with the Land Access Issue Group to discuss areas of overlapping concerns. A majority of the participants agreed to a majority of the recommendations.

The group agreed to the principles that land-use decisions should be made in the context of integrated planning. Planning, environmental assessment, and environmental-effects monitoring should be conducted on an ecosystem basis. It was recommended that all governments design mineral development policies that will be subject to regular reviews and that should reflect Canada's international environmental commitments. Furthermore, it was noted that public involvement mechanisms should be established as early as possible by proponents to reduce delays in the environmental assessment process. Governments should designate a lead agency to manage the process and develop clear criteria for the process. It was also suggested that environmental assessment processes be followed up by multi-stakeholder public liaison committees and regional land-use bodies. Environmental assessment processes should also be established in a way that incorporates the interests of Aboriginal communities. These processes should be consistent with the values, traditions, and priorities of the local communities, as well as with national environmental standards.

During mine operations, it was suggested that the scientific assessment of ecological effects should be based on the best available technology that is commercially proven and affordable. Standards should be appropriate for the specific site. A comprehensive and consistent system for environmental monitoring, which considers cumulative effects and assesses the adequacy of the control systems in protecting ecosystem health, was suggested. Energy efficiency and water conservation should also be considered in the environmental assessment process. The government should support and encourage proactive environmental initiatives by the industry.

Mine reclamation is an ongoing program to restore the ecosystem disturbed by mining. The main objective is to limit long-term environmental impact. The issue of financial assurance to cover the costs of closure and industry liability, as discussed by the Finance and Taxation Group, was also raised in the context of the Environment Issue Group. The ultimate goal is to ensure that adequate funds are available to ensure that disturbed sites may be returned to self-sustaining ecosystems.

Public involvement in decision-making processes may do much to raise the level of trust in the industry and the government, as well as to ensure that necessary environmental standards are met. It may also mean that decision-making processes become interminable and ineffective if the mechanism for including the public is ill-conceived and poorly implemented. Funding for public participation is also an issue that must be considered. Nevertheless, the group encouraged the formation of a well-informed Public Liaison Committee (PLC) that would have the opportunity to review mining activities. There were a number of concerns raised about the composition, funding, and mandate of the PLCs. All of these issues would require careful consideration.

The provision of sound, unbiased, credible, and widely available information about the environmental impacts of a mine was considered an important principle by the issue group. Governments could work toward establishing and standardizing data collection and analysis. A register of qualified scientists and other professionals who could serve as objective participants in an environmental assessment could be established. Furthermore, an interdisciplinary network in land restoration and ecosystem renewal could also be set up. Finally, it was pointed out that costly and unnecessary overlap and duplication in the regulatory process and multiple environmental assessments are counterproductive. It also makes it difficult for the public to acquire information about a project. In particular, federal and provincial environmental standards should be harmonized. A federal facilitator could be appointed to ensure coordination of the regulatory activities of the various federal departments.[15]

The Process of Decision-Making

It is easy to agree with one public official who observed that one 'cannot have much more opposing views than miners and the environmentalists.' The public official observed, however, that the mining representatives were chosen very well; they were diplomatic, in that they were able to avoid rising to the 'baits' some environmentalists threw to them. That perception was echoed using slightly different images by another participant who was interviewed, an environmentalist, Irene Novaczek: 'These were the pretty faces of the mining industry. They were the more progressive members ... and were not reflective of their constituency. One could talk with these guys ... just knowing there were people of that calibre doing something from the inside' was significant in forging new relationships.[16] They were also, however, high-powered and influential members of MAC. It may be that working together revealed that industry seeks to continue to be a credible player in the Canadian economy, contributing to the country's socioeconomic health.

Probably surprising to most environmentalists is the realization that mining is an industry that is, for the most part, willing and able to operate in environmentally friendly ways.

During the first meeting of the Environment Issue Group, few participants were clear about what they were doing there. A government participant noted that one of the reasons progress was possible in the WMI is that people were invited to the table to 'talk more in the abstract, which is much less confrontational than discussions that take place with the pressure of a project going ahead ... It allowed for much more open discussion. Before the WMI, we would deal with Labour with the background of a contract situation or someone's job on the line.'[17]

The environment group had some tough issues to tackle, and as such the process was not smooth. The diversity and complexity of interests around the table were, at times, daunting. Nevertheless, as the list of recommendations demonstrates, a measure of understanding and agreement was achieved and the outcome could be deemed successful. This was no mean feat.

The Workplace/Workforce/Community Issue Group

The Issues and Recommendations

The Workplace/Workforce/Community Issue Group was charged with 'examining the wide range of human resource and industrial relations' questions both across the sectors and in terms of workplace needs.'[18] The group included representatives from the United Steelworkers of America, the Departments of Natural Resources and Human Resources Canada, MAC, Noranda, Placer Dome, and the Canadian Council for Aboriginal Business, among others. They held the first meeting in Toronto on 8 September 1993. At that first meeting four major issues were identified that needed to be addressed: (1) the stability of mine communities in an unstable global economy; (2) standardizing government regulations in a time of declining consensus; (3) integrating Aboriginal workers and their communities into a modern, local mining economy; and (4) employee involvement in mining production planning and the mine-development processes, and the implications for productivity and job security.

The work was divided into three sections: standardization of regulations, community stability, and Aboriginal participation in the mining industry. Under standardization of regulations, a lack of national occupational standards was considered a general problem, as was the lack of uniformity in training and certification programs throughout the country. Further, limited access to appropriate training and apprenticeship programs was identified as an area that could be improved. In general, the mining industry, workers, and communities will benefit if the workforce is well trained and possesses portable skills. To that end, it was pointed out that national occupational analysis could be improved to establish appropriate occupational standards interprovincially and nationally. Industry and educational institutions could also be encouraged to work together more closely to ensure that workers have relevant skills. Although much has been achieved in the area of health and safety in recent years, the issue group noted that this area demands ongoing improvement, particularly as technology changes.

Again, a core set of national standards based on 'best practices' was recommended. It was suggested, however, that the process should have some built-in flexibility for site-specific considerations.

As discussed in Chapter 3, mining communities face many challenges posed by declining mineral reserves, an unpredictable world market, and trends toward 'fly-over' mining. It was recommended that no new single-industry communities should be built unless there is significant potential for long-term socioeconomic benefits for the town. If a mine is to close, residents should be informed well in advance and their interests (including possibilities for community economic development) should be considered in the closure plans. Governments can play an active role in supporting programs for community adjustment and diversification. When long-distance commuting is used for new projects, the industry can be encouraged to draw its labour resources from nearby communities. Other recommendations dealt with taxation and unemployment insurance issues that could be adjusted to accommodate the unique problems caused by mine closure.

It was also noted that Aboriginal participation in mining should be encouraged. Governments could sponsor joint comprehensive employment and training plans to enable Aboriginal people to develop skills for employment in mining. Mining companies could actively recruit Aboriginal students as part of their summer hiring policies. Community economic development opportunities related to mining could also be explored and fostered. Operating mines could develop an employment position of Aboriginal liaison to deal with cultural issues related to employment and training. Aboriginal people should also have opportunities to participate at a more senior management level in the mineral industry. Companies should be encouraged to promote suitably qualified Aboriginal peoples to such a position. Any mining operations that affect or involve Native communities should have the benefit of cross-cultural awareness training, which could be offered to all mine employees. A general mining education program could be offered on a cooperative basis to the community should it so desire.[19]

The Process of Decision-Making

As with the other groups, this issue group prepared a work plan. Background preparation was delegated to different members. The group was made up of strong speakers who were also strong-willed. The group was characterized, in the words of one participant, as both 'frank and polite.' Points were made forcefully, but in a balanced way.

An important 'breakthrough' for the group came when they stopped talking *at* each other. Dan Johnston was hired as facilitator and he began working with the group at their third meeting in December 1993. Under Johnston's guidance, the group agreed to tackle the work using the following strategy:

(1) Listen to the sub-group presentations, then open up the floor for questions.
(2) Discuss the specific objectives of the sub-group.
(3) Explore the issues and impediments blocking the sub-group.
(4) Identify options/recommendations to address issues and impediments.
(5) Think about implementation.
(6) Redraft – for the next meeting – presentations into a common format.[20]

In addition, Johnston offered some background information about how groups such as theirs could develop a strategy. He sorted out what the group wanted to achieve, where it was in December, what it had yet to address, and how it could fulfil its vision and reach its objectives. In the words of one group member, 'Dan slapped everyone between the ears and said, "Let's get going." He brought a bit of sense to the thing.' The participant noted, 'For the first few meetings, I was dumbfounded that so little progress could be made and that at the end of each day it seemed that nothing practical happened. After Dan came, something came out of the day – every day. As the process came to a close, we seemed to develop mutual respect even though it was a diverse group. At the end of the day, you have to take the thing behind closed doors, come to a solution, and address it.'[21]

Issue Group Summary

The work of the issue groups culminated in their reports, which they submitted in the spring of 1994 to the leadership council. The issue groups accomplished three very important tasks in their work: the reports themselves regarding new directions for mining, the forging of new relationships between stakeholders, and improved communications and mutual understanding between diverse constituencies. The formulation of some very carefully considered recommendations have provided a good starting place for public policy-makers to tackle some points of contention between different policy communities. The reports of the issue groups are comprehensive and balanced, incorporating a variety of diverse perspectives. The above brief review of their work does not do justice to their content. This work, in and of itself, merits careful exploration in any future examination of mineral policy in Canada.

The emergence of a different kind of relationship between many of the participants was also an important step toward an improved policy environment. In getting to know each other in the first two meetings, time had been made for the stakeholders to present their own positions. These initial statements of principle were important. It was easy, however, to get mired in principles while failing to develop some pragmatic approaches for dealing with the problems. The challenge was to get over the initial descriptions of stakeholders' perceptions of the problems and to figure out how to move onto common ground. To be sure, some of the discussions confirmed the negative stereotypes various individuals held about each other. Much to the surprise of many participants, however, a fair bit of common ground was discovered. This established an important precedent for future working relationships.

In terms of representation at the meetings, deputy ministers, assistant deputy ministers, and managers would attend. The environment sector sent some representatives who were well known and influential. Industry sent its senior people who brought their clout with them to the table. As in other sectors, of course, personalities and their approaches varied. For example, several industry representatives were able to shatter some stereotypical images. One observer representing the environmental perspective commented that the environment VPs of the mining companies were hikers, 'green VPs who sang from the same song book.' Progressive and knowledgable environmentalists who had previous exposure to the mineral industry also attended, and shared the concerns of the industry. They were able to garner the respect of some members of the industry.

The Environment Issue Group may have been the group that worked hardest to develop social cohesion. The group decided that all members should contribute to writing reports. It divided its members and the issues to be tackled into three sub-groups based on the mining process itself— the pre-mining, operational, and mine-closure stages. Each sub-group had six to eight members. It was the first group that faced serious challenges to reaching consensus, since it is the pre-mining stage that is profoundly defined by environmental assessment processes. The mine-closure subgroup, by contrast, was much more efficient, namely because there were fewer and less complicated issues with which to grapple. Of course, personalities varied significantly; there were those people who had abrasive personalities, those who were prepared to believe the worst about other policy communities, and then, there were those who were particularly skilled at interpersonal relationships. The sub-groups themselves were fairly cohesive, but there was less cohesion among the greater group. This was one group, according to a group member, that should have had more than one Aboriginal representative.

These differences in group dynamics influence the policy field itself. The nature of the issues, personalities, past relationships and experiences, information and knowledge competencies and gaps, and the abilities of the chairperson are among the early determinants of group dynamics in a consensus-building process. There were certainly some obstacles to overcome. From Irene Novaczek's perspective, 'we, as environmentalists, were there as bowling balls that government could throw at industry ... As environmentalists we often hope and trust that some departments will see us as their natural allies, but often we get the feeling that's not the way we're being viewed. Instead, we're handy bowling balls. Governments let us do their dirty work, then they don't have to say what their motives are.'[22] From one public official's perspective, however, it was government that sat in the tough seat; he perceived that the WMI was seen as 'government's responsibility.' Several interviewees from different sectors observed that government had a big learning curve to climb. One of the obstacles in the process is that some public officials assume a rather passive position, and will not, for example, come forward with their information and perspectives.

There are two keys to success that stand out in the formative stage of group processes: building understanding and gaining trust. Art Ball reminds us that 'understanding is the result of trust, and trust is the result of understanding.'[23] He observed that the transformation in trust and understanding in the WMI process was amazing. Of course, there were some individuals who remained entrenched in their original perspectives and positions. To combat stereotypes and misinformation or the lack of information, it is necessary to ensure that well-researched, timely, and accurate background factual material and expert testimony (an 'expert' is defined broadly to include those who bring valuable traditional knowledge) are available to group members. It is clear that because it was uncharted terrain, the WMI participants were suspicious of the process. Dan Johnston observed: 'People did not want to be seen to be reluctant to be there – to give the impression that they were not trying. They were sceptical but when they started to reach agreement, they began to be pleased about the process.'[24]

In their first meetings, group members may have sensed that their new colleagues – who were sometimes old antagonists – were 'holding something back' and creating 'smoke and mirrors' to gain advantage, whether it was government, environmentalists, or industry. For example, one industry participant observed: 'It was difficult to trust some of the representatives from the environmental groups. They weren't straightforward, but there were some really super environmentalists that had a lot of integrity. It was those people who convinced us that there was integrity on the other side who wanted to make this work.'

Within every group, within every sector, there will always be participants who make it difficult to forge new relationships and set aside myths and misinformation. Significantly, however, the WMI revealed that it only takes one or two participants 'on the other side' to make a significant, positive contribution. These are the individuals who offer some indication that they do not fit the negative stereotypes that persist and they are willing to construct new relations built on consensus.

Moving toward mutual understanding was crucial to achieving some minimum degree of consensus. The process was plagued by a number of problems. The terminology used was subject to numerous competing interpretations. The addition of new participants meant that processes were set back as group dynamics changed and the new participants were introduced to the work-in-progress. Concerns were raised about overlapping issues dealt with by the groups. Several more meetings were scheduled than initially expected. When the leadership council began to consider the work of the issue groups, some members of the latter groups felt that their ideas and concerns were not being adequately addressed.

It is one thing to share information, but finding common meaning in the flood of information received – from the experts who gave group presentations to the reports and papers that were faxed across the country to elucidate and educate – depended on achieving a condition of shared knowledge. Participants needed to get to a stage in which they understood the important concepts particular to each of the five sectors, shared a common language usage (a challenge given the scope of technical expertise and specialized knowledge), and shared interpretations about the problems they identified. At a Land Access Issue Group meeting, efforts were made to define terms such as 'option value,' 'inter-generational bequest value,' 'futility rate,' and 'non-transferable qualities.'[25] Questions had to be asked, such as: What is the social value of metals? What is the monetary value of intangibles? Debates about language usage, conceptual clarity, and gaps in meaning, measured by the diversity of the stakeholders around the issue groups' tables, were necessary.

Examples abound in the WMI about the importance of language usage to a diverse group that had been challenged to reach consensus. The minutes of the Land Access Issue Group include questions such as: Is it government 'interference,' or would 'intervention' be a better word? The Environment Issue Group, as with the other groups, worked hard to find precise meanings that all members could find acceptable. The groups' exercises were studies in precision. For example, it was suggested that the words 'are involved in decision-making processes' should be changed to 'participate in decision-making.'[26] Another example involves an effort to refine a draft principle that read as follows: 'Developing environmentally responsible mining operations requires a preplanning process that addresses broad concerns such as the threat to biodiversity and ecological integrity.'[27] To reach consensus on meaning and wording such as the above reflects hours of hard work and frustration. In the process of tailoring it further, the following analysis was made within the Environment Issue Group:

> A variety of concerns were expressed about this principle. It was suggested that negative language such as 'threat' should be avoided. It was also suggested that if technical language such as 'biodiversity' or 'ecological integrity' is used, a glossary should be created to provide definitions or

> explanations. This suggestion applied to all the issue groups. It was also suggested that these terms are uncertain and evolving and should therefore not be used.[28]

Clearly these quotations reflect depth of understanding, precision of wordsmithing, and sensitivity, which is an accomplishment not to be undervalued. For example, concerns were raised with using the term 'less than 5 per cent of Canada is set aside.' It was noted that the phrase 'less than' was 'biased and infers a paltry amount.' Replace 'less than' with 'about' or 'nearly.'[29] The concepts were so 'loaded' that even in May 1994, when the issue groups were struggling to conclude their reports under heavy time constraints, requests were made to define terms such as 'ecological integrity.'

Dan Johnston, who joined the WMI as a facilitator for an issue group midway through their work, and later became a facilitator in the leadership council from May 1994 until the signing of the accord in September 1994, stated: 'To a large extent, people started using terms everyone else did. Issue groups had an even better shared understanding because they spent more time together.'[30] Without the ability of the issue groups to arrive at a shared understanding, the WMI would have foundered. It has to be remembered that for most of the participants, if not all of them, their WMI work was in addition to their regular workplace responsibilities. George Miller emphasizes that the 'Issue groups owned their reports ... They'd fought over those words, and they knew how delicately balanced consensus was. Once they agreed to the words, they fell in love with them. They knew how important every word was!'[31] Given their relatively short work period – approximately nine months between their first contact and their submission of draft reports – the intensity of their efforts and the level of achievement is obvious.

The issue group reports were produced despite challenges such as changes in group membership. Continuity of members' participation is crucial when a process depends on nurturing trust relationships, increasing shared understanding, and creating a growing list of agreed-upon issues and recommendations. When a group's dynamics are settling into a more comfortable, more predictable, established pattern in which positions are formulating and agreements are being struck, the introduction of a new member – with his or her own assumptions and values, and without the benefits of the group's learning curve established in the previous three or four months – can set back the work. One labour representative agreed that a change in cast of characters, as experienced in the Land Access Issue Group and in other groups, 'changes the way people can contribute to meetings. It changes the dynamics.'[32] Continuity of membership, then, is an important factor for success.

The issue groups met more often than they thought they would. The

Land Access group had six meetings, Environment met five times, and the other groups met six to eight times. The meetings were typically two-day events. The groups met across Canada, on the basis that a manifestation of the 'national' process was important; the exception was the Finance group, with its corporate culture, which never left Toronto. The March 1994 meeting of the leadership council was set to give the issue groups a chance to present briefings of their work.

Annual Meeting of the Mines Ministers: September 1993

While the issue groups were holding their first meetings at different times in the summer and fall of 1993, the WMI secretariat was providing information and communication links between the issue groups, the leadership council, and the working group. In September 1993, a meeting of mines ministers was held in New Brunswick. By that time, the working group had a sense of mission. It had been meeting at monthly intervals, and member of the group expected that their involvement would diminish as the issue groups started their work. Some of the most active participants left the working group, resulting in a loss of important analytical and other skills. The leadership council was not yet called to its work. In October 1993, Dixon Thompson submitted his remarks about the possible roles that the leadership council could assume to the working group. He suggested that the council could:

- provide advice or suggestions to the issue groups and the working group on process, strategies, and substance
- learn about the matters of concern to the issue groups as early as possible because of the commitment to communication (usually a two-way process) and because of the desire to avoid surprises late in the process
- establish effective two-way communication with the issue groups with an understanding that at some stage it will be necessary to respond, possibly with something more than words
- receive, review, and revise reports for the issue groups; the process should be arranged, if possible, so the issue groups can respond to leadership council comments and criticisms
- develop and oversee the processes for implementation.[33]

As described above, in the summer of 1993 the fundamental issues were defined by the issue groups, time-frames were established, and process requirements were somewhat settled. At the September 1993 ministers' meeting, however, there was a perception among the mines ministers that nothing had been done, and they were becoming uncomfortable in their support of the project. Of course, temporal perceptions in the realm of politics tend to run along four-year electoral cycles. When those in political office approach the second half of their terms, the need intensifies to have tangible, positive products to show the electorate.

Contrary to the perception that little was being accomplished by the issue groups, there had, in fact, been significant progress. George Miller revealed the budget shortfall and circulated a memo stating that he was confident new funding sources would be found. Some governments, such as Quebec's, *were* positive about the process. Although Quebec was not participating in the leadership council nor the working group (though they were involved in the issue groups), the province was supportive of the WMI and had paid its contribution. Ministers from the Atlantic provinces were most supportive; Nova Scotia's minister, Don Downe, observed that, given its economy and its small land mass, mining is even more important to Nova Scotia than it is to Ontario, the largest mining province. The government of Alberta chose not to participate and the role of Saskatchewan was not one of a fully enthusiastic participant. One reason for this lack of participation and reluctant involvement is that the Mining Association of Canada, which initiated the WMI, is an association that represents the metal industry. As mentioned earlier, the Alberta government, whose mineral wealth primarily resides in oil and coal, saw no reason to participate. Similarly, Saskatchewan does not have a large metal-mining sector; representation in the WMI was based on the uranium and potash sectors, and their companies have different associations to which they belong, not the MAC. Dan McFadyen, the ADM in Saskatchewan, stated: 'We've a large mining sector in Saskatchewan ... but none of Saskatchewan's mining companies are members of MAC.'[34] Throughout the WMI, the Mining Association of Canada came to be viewed by some as an umbrella organization for the mining industry, although it only represented a particular domain within the mining industry; 'it got labelled nationally as representing Canada's mining industry.' This happened despite the fact that an umbrella organization was created after the WMI process was initiated in 1992 – the Canadian Mineral Industry Federation. In addition to such provincial differences in mining operations, the WMI was not enthusiastically supported by all mines ministers at their meeting in September 1993. This new labour-intensive policy exercise was not something ministers could readily adopt, given the multiple demands on their time. Despite governments' financial concerns, pressures for a product from some politicians, and other considerations, the process continued to evolve and move along.

Communications and Implementation Committee

A Communications and Implementation Committee was formed in December 1993, bringing in people from every level – the leadership council, the issue groups, and the working group. It was to be the committee that would integrate the work and concerns of the three groups. George Miller observes: 'Toward the end of the process, it was found necessary to create a cross-cutting body, the "Communications and Implementation Committee," with membership from all three levels – the leadership council, the working group, and the issue groups. This committee thought strategically and politically about what kind of an accord the leaders could endorse, how the results could best be communicated, and how the good will and momentum could be maintained after the consultations finished. It also eventually became a drafting committee for the Leadership Accord itself and assisted in resolving tough issues in the closing days.'[35] At the December meeting, the Communications and Implementation Committee discussed the elements of an implementation strategy.

The Leadership Council Meetings

The March 1994 leadership council meeting was the group's first comprehensive meeting. Briefing summaries of the issue groups' work to date, pulled together by the secretariat, were presented. More ministers attended the March 1994 meeting (seven) because the meeting coincided with the Prospectors and Developers Association's annual convention. As well, many more environmentalists were represented.

At the second comprehensive leadership council meeting in May 1994, many new participants arrived from the environmental and Aboriginal sectors; they faced all the challenges of not having shared the learning curve gained over past months' efforts. The addition of new participants around the table posed a problem; these people had not worked through any aspect of the process before, and they had not been part of the learning curve that other members had experienced. Over the previous year, issue group members were adopting strategies about how to face conflict constructively. The consensus-building model that evolved in the WMI ran counter to learned and, perhaps, intuitive behaviour.

The May 1994 meeting was the first time the leadership council really had to work together. At the meeting there were approximately forty-four people around the table and only a handful of observers. Dan Johnston was called in to assist as co-facilitator. Johnston was a good choice. He led the meetings with an assertive style. When people could not find the right words, he offered suggestions. In hard policy areas, a facilitator sometimes is able to push in a way that one would not have to in soft policy areas, such as social policy. Johnston's assertiveness particularly appealed to the industry leaders; in the words of one observer, they appreciated his 'edge.' One participant watched Dan Johnston make a 'B-line' during coffee breaks, trying to '*do a bilateral* over coffee between two opposing or contesting sides.'

The co-chairs of the meeting were BC Mines Minister Anne Edwards, George Connell, and Dan Johnston. George Connell stated that the purpose of the meeting was to determine 'whether what's in the briefing books is sufficient for packaging policy ... [is it] the foundation for the mining industry of Canada, as we envision it in the future?' The goal of the meeting was to answer: 'Is it a satisfactory package or do we need to figure out how to make it that way?' MAC president, George Miller, expressed the need to 'agree on a charter, a set of principles which represent a consensus of all senior representatives of the groups involved.' The objective was to work out a consensus about the principles that would be set out in the accord. Broad principles would be the heart of the accord, putting aside the many recommendations produced by the issue groups. Some members were left wondering whether anything substantive would be reflected in the accord.

As noted earlier, the documentation presented to participants at the May meeting was ordered along 'principles and objectives,' an approach to which some people objected. Comments from issue group representatives on the leadership council were expressed at the outset of the meeting, such as 'who re-interpreted this?' and 'what's happened to our work?' The concerns were based on structural and procedural questions, and they reflected a high level of discomfort with the evolution of roles and responsibilities. There was a lack of common understanding among all the WMI participants about the role of each of the groups. For example, was the work of the issue groups to serve only as advice for the leadership council? What would be the status of the completed issue group reports?[36]

It became clear during that meeting that 'total harmony between Issue Groups and Leadership Council on principles and objectives may not be possible.'[37] In the end, the suggested action plan of the WMI leadership council assumed that:

- it is highly desirable that the principles, objectives, etc., contained in the WMI Leadership Accord are harmonized, to the greatest extent possible, with those set forth in the Issue Group Reports
- however, the leadership council is not bound by the Issue Group Reports (i.e., they are treated as advice or recommendation to the leadership council).[38]

The lack of clarity clearly frustrated the issue groups, who perceived that their work was being diluted. The leadership council went through the briefing book painstakingly, line by line, word by word. Language usage was very important. There were also some past negative encounters between participants that adversely influenced the process, and it would not have mattered if they were only talking about where to place a preposition; there would still have been disagreement.

On the positive side, important relationships were being developed, renegotiated, and, sometimes, recast in a more positive light. Networks were forming between sectors and within sectors. Governments themselves were developing important contacts across the country. The process was facilitating the establishment of communications networks and ideas about how the same problems may be tackled more efficiently.

Expectations about the process varied tremendously. Some people focused on the product that was due on 13 September 1994, perhaps thinking that it would be 'the answer' to the problems that they perceived. A few others, like George Patterson, walked into the process with the view that the WMI process was only 'one of a number of vehicles ... it was not *the* answer.' It was the process itself that was important – the process that facilitated dialogue, trust, and understanding.

The industry found itself at a May 1994 leadership council meeting in which the dynamics were vastly different from the 'chat' with stakeholders that had been anticipated when the WMI was conceived. Similarly, the process presented unfamiliar challenges to government representatives; mines ministers are rarely, if ever, involved in the kind of plodding work undertaken at the May meeting, where each word was assessed, line by line. The expectations participants brought to the table affected the process; for example, Aboriginal representatives in the issue groups saw themselves as government leaders and queried why they were dealing with the ADMs and technical people sent by provincial governments. The brief encounter of the May leadership council meeting was constraining to say the least – players who had not worked together in a consensus-making forum were expected to cover this complicated subject peppered with value-laden terminology in two days!

When asked about time constraints, the responses were diverse. An environmentalist stated that time was a very frustrating constraint: 'It was clear there were no more dollars and that there would be no further meetings. Things were cut off just as we were getting going. As interesting as it all was, as participants, we have to ask whether the taxpayer was getting optimal value for his/her dollar. We'd often look at ourselves and think that we had so much to do at home.' On the other hand, a facilitator concluded that 'The debate was not cut off prematurely. At the end of the day, there was a document.' A mines minister noted: 'The input is needed and I am enthusiastic about the process, but one can't sit forever, everything has its time, however. WMI had its time and right now we must play it out on the provincial tables ...'

Several people interviewed thought that the time was constrained but that it would be difficult to carry on any longer with the process. An industry representative reflected: 'To arrive at Agenda 21 in Rio '92, four preparatory international conferences were staged ... a six-year enterprise. One almost needs that kind of time-frame to realize such depth ... All our negotiations were, in one way or another, the work of volunteers. It was an 'on top' job for all participants. We were all fatigued and could not have sustained the process ... If the stakes are high enough and everyone is working on it full-time, one could sustain the process longer.' An assistant deputy minister noted: 'We were rushed ... The process was slow getting started. I don't think anyone literally assumed the leadership. No one took the bull by the horns. We lost a good part of the first year. I don't think two years is too short, but with the slow start, everything was condensed.' Another assistant deputy minister agreed and said 'Yes, time was too constraining.' One government observer noted that 'A lot of people thought that the process dragged on, including industry. It was a fairly expensive process at the end of the day, and in days of difficult fiscal situation, the issue of financing loomed. There was pressure for a product.' Finally, an Aboriginal representative noted that the process was 'very rushed. It was trying to change 100 years of business.'[39]

Enormous effort and commitment was expended by those who participated. Further, no one would disagree that there were heavy time constraints, making it even more impressive that so much conceptual, technical, and procedural territory was tackled by the issue groups, and then to a different degree, by the leadership council. Indeed, given the fluid, unstructured WMI process, the signing of the accord in September 1994 was a remarkable achievement. Nova Scotia's minister, Don Downe, states that it 'was a tough process. There was a certain amount of water in everyone's wine ... that was necessary to come up with a solution.' He stated that there was enough political will to sustain the process and that Nova Scotia 'put as much political capital as possible' into the WMI.[40]

Not everyone was as impressed. One First Nations representative reflected that being a 'representative of a community or a grassroots component leads to more fundamental dialogue. The process became more cleansed as it filtered through the working group, and it became more sterile at the leadership council level. There were no reality checks to maintain the integrity of depth at the working group and leadership council level; they were disconnected.' Another First Nations representative maintained that the leadership council was 'purely a political thing with its own mandate ... The WMI was nice, like going for a ride at Canada's Wonderland.' From a different vantage point and in retrospect, one participant said that the leadership council was needed to go through the issue groups' reports and come out with a single statement.

Even if the implementation process is frustratingly slow to some, the outcome of the WMI process must not be measured in only observable and quantifiable terms; for example, by having sat together and shared words and ideas, future relations between representatives of various sector groups may be *qualitatively* different.

For the first time, in May 1994, fundamental design questions were addressed. It was observed in the *final* few minutes of the May meeting: 'This is the third Leadership Council we've raised this ... we've still in part failed to communicate from the Leadership Council to the Issue Groups what the Leadership Council roles and responsibilities are and the relationship between the two groups ...'[41] Similarly, the issue of implementation of the recommendations to be offered by the WMI leadership council received little consideration until the last months of the process. The secretariat prepared several alternative implementation strategies that were included in the briefing book given to the May participants. The process issues included the following: How would the leadership council and the issue groups move from the May meeting through to their deadline of September? What implementation process would be set in place after the submission of the accord in September 1994? The time allotted to address these questions at the May meeting was too brief. Figure 5.1 illustrates one perspective of the flow of proposed recommendations, commitments, and undertakings through the WMI process.

The challenge of confronting 'language as politics' took all the time of the meeting and there was not time, formally at least, to discuss implementation. Communication strategies and implementation plans were left to the Communications and Implementation Committee, which had the

Figure 5.1

WMI outputs: Concepts

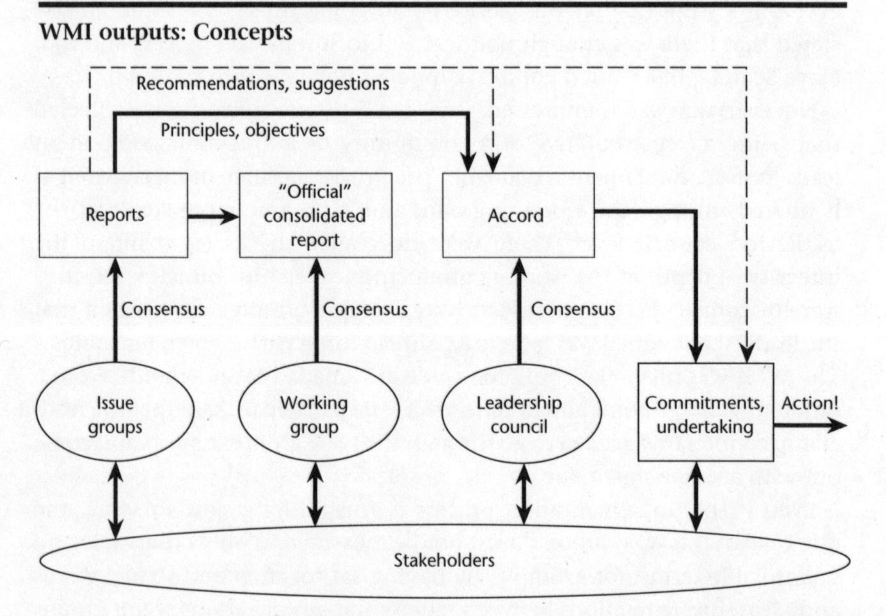

Source: Whitehorse Mining Initiative, Secretariat's Office, 11 May, 1994.

impediment of working with a limited budget. A public-relations firm was approached over the summer to see what outreach communication strategy could be undertaken after the submission of the accord on 13 September.[42]

Participants seemed reluctant to address the subject of implementation until September 1994, since no one was sure there would be anything to implement. Most pressing on everyone's minds was getting a document ready, and agreed upon by the leadership council, for the mines ministers' meeting in September.

It was up to the Communications and Implementation Committee to develop the leadership council accord in time for a July meeting. They decided that since the process had two tracks, there would be two documents. Rather than *summarizing* the work of the issue groups, the leadership council accord would be separately developed. Placing the issue group reports in the accord as appendixes gave them the recognition that leadership council members agreed they deserved. This also dealt with one industry member's concern of explicitly endorsing the recommendations. Federal minister Ann McLellan stated that such a two-track approach gave adequate recognition to the importance of the issue groups as a 'rich source of inspiration ... but federal government will not sign on and endorse [those reports] either implicitly or explicitly.' By the July meeting, there was a draft accord that had been written with the help of a professional writer. The main issue, which remained unresolved after the July leadership council meeting, centred on whether the accord should include clearly articulated financial assurances for mine reclamation.

The July 1994 leadership council meeting was well attended. Time was a severe limitation – one day and a half to assess and revise the draft accord. It was noted that international attention was on the WMI process, which was a good comment to make at the outset; it reminded people of the distinctiveness of the process to which they were contributing and of the importance of achieving consensus so there would be something to deliver in September.

Champions are crucial to the success of a process like the WMI. George Miller, the central figure in the WMI process, did not attend the July meeting. People were heard saying, on the first day, that they wished George could be at the meeting. Walter Segsworth of Westmin Resources, however, stepped in effectively. The details covered the whole spectrum of issues, from consideration of environmentally responsible tax regimes to protected areas. As at earlier meetings, it was a small group that consistently tackled the issues. A valuable contribution was made by some members of the environmental community in offering alternative interpretations or wording. A representative from Alberta spoke up at the end of the meeting with a brief consideration of implementation; he said he spoke with his deputy minister, and the decision was to 'stand aside from these discussions, since Alberta was not involved in drafting the Accord.' In this case, the concern was that many issues went beyond the jurisdiction of the Alberta minister of Energy and Natural Resources.

Discussions about the WMI 'vision statement' revealed that participants continued to have trouble with what was meant by the terms 'principles' and 'objectives.' Dan Johnston asked the group to 'agree on terminology of what is now called shared assumptions and goals, and which were previously called principles and objectives.' Participants who offered critiques were encouraged to draft alternatives during the breaks in the meeting. Given this prodding, the actors in the leadership council became more active participants and motivated toward enlightened self-interest. In a very intense, condensed time-frame, they appeared to be increasingly willing to participate in collaborative thinking.

It was the land-access issues that presented the most pressing problems. In particular, the concept of 'protected areas' posed a serious challenge in negotiating consensus. A sub-committee was formed after the July 1994 meeting to deal with the major unresolved issue: finding consensus about the degree of comprehensiveness that should be applied to the concept of 'protected areas.' Debate over the issue did not revolve around the language itself. Instead, the trouble boiled from the past relations between particular actors. Throughout the summer of 1994, Johnston made sure that the WMI participants stayed in communication, using telephones to link active participants across the country. George Miller stated that he 'was astonished that so much could be accomplished around conference calls.'[43]

The Whitehorse Mining Accord

On 13 September 1994, an accord was signed – something many participants would not have predicted at the beginning of the process and much to the surprise of many. The substantive themes addressed in the accord were present in the work of the issue groups as outlined above. The accord included a vision statement (see Appendix B), principles and goals, and a statement of commitment. The primary themes were broken down into areas of challenges:

Addressing Business Needs: Business Climate, Financing Taxation, Overlap and Duplication, Government Services

Maintaining a Healthy Environment: Environmental Protection, Planning and Environmental Assessment, Use of Information and Science in Environmental Decision-making

Resolving Land-Use Issues: Land Use and Land Access, Protected Areas, Certainty of Mineral Tenure

Ensuring the Welfare of Workers and Communities: Attracting and Retaining Skilled Workers, Maximizing Community Benefits from Mining

Meeting Aboriginal Concerns: Aboriginal Lands and Resources, Aboriginal Involvement in the Mining Sector

Improving Decisions: Open Decision-making Processes.[44]

One principle in particular – Protected Areas – serves to illustrate the degree of consensus that had been achieved. It stated, 'Protected area networks are essential contributors to environmental health, biological diversity, and ecological processes, as well as being a fundamental part of the sustainable balance of society, economy, and environment.'[45] It is understood that there is a significant difference between arriving at a principle and subsequently putting the goal into practice. Nevertheless, an ability to achieve a consensus on this particular issue (as well as many other controversial subjects) illustrates the commitment of the participants to arrive at an accord. Most of the participants interviewed, particularly those in industry, stated that they went much further in their understanding of other world views and in their willingness to compromise than they ever expected. Industry's major gain from the WMI lies in its higher credibility in the eyes of the stakeholders, and the dividends of such changed perceptions are likely to continue in their relations. What should not be undervalued is the importance of the process itself and all the subtle changes in understanding of language, concepts, ideas, values, objectives, and personalities.

Finally, a signed statement of commitment was included, which committed signatories to pursuing the goals and principles outlined in the accord:

Whitehorse Mining Initiative Commitment

It is essential that we translate **Our Principles and Goals** into action, and that we maintain a framework for an ongoing relationship.

Therefore, we, the undersigned members of the Leadership Council, undertake to:

- promote the Accord within our respective constituencies and familiarize constituents with the Whitehorse Mining Initiative recommendations
- support its Vision and advance its Principles and Goals, in the interests of all Canadians
- in cooperation with other stakeholders, develop and adopt action plans to give effect to the Accord within our respective jurisdictions

- support and encourage individual stakeholders who wish to undertake actions consistent with the spirit and intent of the Accord.

> **We, the members of the Leadership Council**, sign and endorse this Accord. We recognize that all WMI stakeholders have an important role in the future of mining in Canada. We undertake, in the interests of all Canadians, to support its Vision and advance its Principles and Goals.[46]

Establishing the mining industry's credibility is no small achievement, in and of itself. What has been accomplished is the construction of some broad parameters in which people feel they can reasonably achieve their goals and make some progress in the future. There was concern that if an implementation plan was not immediately apparent, it would have been very easy to dismiss the process. Such an emphasis undervalued what occurred in the time period leading up to the final accord.

In the WMI process, Aboriginal peoples and labour and environment representatives showed themselves to be adept in developing strategies and in using the resources and other skills of policy analysts and lawyers. Many participants from industry had a background in mining and geology and were not as adept at defining and articulating their needs, nor, perhaps, the significance of gaining an understanding of others' interests and values. Various participants commented that the WMI was only a beginning step, that the relationships should be pursued, and that the results should be communicated to the public and to stakeholders' constituencies in a concerted, strategic communication exercise.

If we look at the set of goals for the success of the WMI laid out at the beginning of Chapter 4, we can see that many of the rules were not followed. There is no question that the process was needed; people from different groups had a desire to meet and work together in a new, more conciliatory fashion. Most of those interviewed thought that the right people had participated and the membership was fairly balanced, in spite of under-representation in some areas. There was certainly a deadline that the participants did observe, and this helped them stay focused. There was not, however, an agreed form for the outcome or a defined plan about what to do with the results. This continues to dog the efforts of those who try to carry forward with the momentum initiated by the Whitehorse Mining Initiative. Periodically, support of key decision-makers waned, particularly when some people left the process and new participants joined. It was the efforts of a few, supported by many, that often defined the ultimate success of the Whitehorse Mining Initiative. With the signing of the accord, an important first step on a very long road has been taken. As the following examination of federal and provincial initiatives after the signing of the WMI illustrates, it is clear that some progress is being made on many fronts. While it is true that many of these activities would have taken place without the WMI, the accord helped establish a policy framework that would allow federal and provincial jurisdictions to move toward a common set of goals. It is also clear that those who signed the statement of commitment have a huge set of challenges ahead.

6

Implementing the Vision? Provincial and Federal Initiatives

The Whitehorse Mining Initiative is unique because of its national, sweeping scope and because it was initiated by industry, not government. It is important to note, however, that it was, in many ways, a product of the times. The 1980s and early 1990s saw a general movement toward 'roundtable approaches' to sustainable development and land-use planning. Such approaches involved the development of partnerships with various professional and resource-user interest groups.

Often, political interest is at its peak when various constituencies can agree that something should be done and they can put pen to paper outlining a vision of the ideal goals for resource management. Such vision statements help guide the making of legislation while providing the elected government with an opportunity to attract positive media coverage. In practice, however, it is implementation and administration that will determine the long-term success of a policy. All too often, when devising policy – which is derived from consensus-based, multi-stakeholder approaches – the scope becomes so large that governments are somewhat at a loss about how to realize its effective implementation.

Different jurisdictions have decided to implement the goals of the WMI in a variety of distinctive ways. A look at a few of the federal and provincial initiatives in the year immediately following the accord highlights the more active efforts at implementation. Specific attention is directed at British Columbia and Ontario, which are large mining provinces that set up their own advisory councils. This chapter is primarily concerned with outlining some of the initial activities that soon followed the signing of the accord and the interpretation of the WMI by provincial officials. Critiques by various stakeholder groups about the WMI and implementation are provided in the concluding chapter.

The Federal Government

The Intergovernmental Working Group on Mining has been charged with overseeing the implementation of the Whitehorse Mining Initiative across Canada. One of the valuable outcomes of the WMI was the framework that it provided to the Mining Sector of Natural Resources Canada for interdepartmental discussions. Natural Resources Canada established a WMI Interdepartmental Committee to 'ensure that the WMI goals are addressed by all federal departments and agencies.'[1] Rather than pursuing its interests in isolation, the issues and concerns related to the achievement of sustainable mining is now much more salient on the agendas of other government departments. In preparing the implementation report of the WMI Accord, Natural Resources Canada (NRCan) consulted with Environment Canada (EC), Canada Heritage-Parks (CH), and others. The Department of Indian Affairs and Northern Development (DIAND) responded separately because it was a co-signatory on the WMI Accord.[2]

The federal government continued to coordinate the work of the WMI Accord throughout 1994-5. NRCan released a discussion paper in September 1995 entitled *Sustainable Development and Minerals and Metals*. The federal minister (NRCan) established an advisory committee on WMI implementation with representatives from non-governmental sectors that participated in the WMI. Consultations with the Advisory Committee is providing the basis for the development of a new federal Minerals and Metals Policy based on goals for sustainable development. Many of the policy announcements presented by the federal government were projects already under way. The WMI provided a structure with which to coordinate the implementation and delivery of these programs.

Until very recently, there have not been many tangible outputs of the WMI introduced by the federal government. There would likely have been even fewer policy outputs had the WMI not raised the mining industry to a higher profile. Bill Toms of Finance Canada states: 'One of the reasons that Mr. Manley selected the mining industry as one of six sectors identified as part of the initiative's regulatory reform component in Industry Canada's report *Building a More Innovative Economy* was as a result of the WMI, which moved mining up the queue.'[3] The Honourable John Manley, minister of Industry Canada, announced the federal initiative on 5 December 1995, and it is one of four components in the federal strategy to foster job creation and economic growth. The regulatory issues that will be addressed by the initiative include:

- administration of the Fisheries Act
- land use and related decision-making
- definition of waste
- permitting process north of 60°
- regulatory impact analysis
- toxic-management policies and practices.[4]

The discussions initiated by the WMI informed federal public officials about the implications of the complex, contradictory regulatory environment in Canada. Federal officials re-examined questions such as whether tax incentives are the right policy answer to regulatory issues. One federal government representative noted that they had to consider whether they were focusing on the real problem. Follow-up action on a number of issues took place on 18 October 1995, which was 'Lobby Day' for the mining industry. The Mining Association of Canada and other groups brought senior executives to Ottawa to meet with members of Parliament. The next day, industry representatives met with NRCan and Environment Canada officials at a Regulatory Streamlining Workshop (sponsored by NRCan and the MAC) 'to go through the phalanx of regulations and to look at what Manley and the WMI talked about.'[5] Three major issues drew their attention: fish-habitat management; environmental assessment; and land-use planning. At the workshop, 'seventy-five participants from federal departments, industry, and industry associations agreed that the responsible federal agency/department and the mining industry should establish a formal mechanism for addressing industry concerns. Particularly during this period of economic restraint, regulatory reform offers us an excellent mechanism for enhancing competitiveness, encouraging investment, and creating jobs.'[6] It is due to the impetus of the WMI and continued lobbying efforts to the federal government that some policy action has been taken.

Implementation of the WMI goals and objectives have fallen across departments, involving Environment Canada (EC), Canada Heritage-Parks (CH), the Department of Indian Affairs and Northern Development (DIAND), and of course, Natural Resources Canada (NRCan). The DIAND submitted a separate progress report to the 23 November 1995 WMI follow-up meeting, due to its territorial responsibility and because it was a co-signatory of the WMI Accord.

All federal ministers, as stated in the *Guide to Green Government,* must now make sustainable development a requirement for their departmental priorities. In January 1995, new NRCan legislation came into force, which binds the minister to uphold the principles of sustainable development in exercising powers and performing duties. In September 1995, the minister of NRCan released an issues discussion paper, 'Sustainable Development and Minerals and Metals.' Consultation with other federal departments and non-governmental stakeholders was being facilitated through the minister's WMI Advisory Committee; responses were used to draft a new federal Minerals and Metals Policy based on the principles of sustainable development. The Advisory Committee was composed of representatives from the non-governmental sectors that participated in the WMI. Another committee, the WMI Interdepartmental Committee, is chaired by NRCan; it was established to ensure that the WMI goals are addressed by all federal departments and agencies. The committee's purpose is to 'identify areas where the federal government is undertaking work to address specific WMI goals' and 'to ensure that the initiative is continually kept in front of other departments and agencies.'[7] Of particular concern to many communities of interest was the issue of harmonizing federal, provincial, and territorial environmental management frameworks. In June 1996 the federal government tabled a response to the House Standing Committee on Natural Resources interim report, *Streamlining Environmental Regulations for Mining*. The government supported recommendations for streamlining regulations while maintaining high standards of environmental protection.

Canada Heritage-Parks has been addressing one of the three WMI goals related to protected areas – to create and set aside from industrial development by the year 2000 those protected areas required to achieve representation of Canada's land-based natural regions; work is progressing toward the completion of a national parks system. There are several important parks that have been established. On 11 July 1995, the federal and BC governments signed the Pacific Marine Heritage Legacy (BC), a protected-area initiative to represent the Strait of Georgia Lowlands Natural Region. A new national park, the Vuntut National Park (Yukon), was negotiated through the Council of Yukon Indians land claim and came into force on 14 February 1995 through a consequential amendment to the National Parks Act; six oil companies relinquished their oil and gas rights within the park area as a goodwill conservation gesture, which was required for the park to be established. Other parks being studied are the Bluenose, Wager Bay, and Northern Bathurst Island in the Northwest Territories. In addition, Environment Canada is leading a number of policy initiatives, involving various stakeholders, that are consistent with the intent of the WMI; the Toxic Substances Management Policy issued in June 1995 and the Pollution Prevention Action Strategy, issued the same month, are examples of the department's efforts. It should be noted that many of these kinds of activities could and do take place independent of comprehensive agreements such as the WMI. The WMI does provide a useful framework, however, to organize policy and serve as a checklist for activities that are taking place, as well as for those that have yet to be implemented.

National Mining Week was announced by Ann McLellan on 5 May 1995 to focus Canadians' attention on the importance of mining as a technologically advanced, essential component of Canada's economy. Complementing this action to increase awareness of the industry's progress and contribution to the economic and environmental health of the country is the Canadian Industry Program for Energy Conservation; seven mining companies were honoured on 12 September 1995 for their 'visible examples of the mining industry's commitment to meet the goal of sustainable development.'[8] Another example of the federal effort to encourage voluntary action undertaken by the industry is the Accelerated Reduction and Elimination of Toxics (ARET) program; thirteen members of the MAC committed themselves to achieving, by the year 2000, a 70 per cent reduction of designated toxic releases. Other collaborative efforts include: the Mine Environment Neutral Drainage (MEND) program, which involves four federal departments (NRCan-CANMET, EC, Fisheries and Oceans, DIAND), eight provincial governments, the mining industry, and others; the Aquatics Effects Technology Evaluation program, a government-industry initiative (three federal departments, eight provinces, and the MAC), which develops 'the base of scientific and technological expertise to provide both industry and governments with more cost-effective methods of assessing the environmental impacts of mining effluents.'[9]

The Department of Indian Affairs and Northern Development has responsibility for managing resources in the Yukon and the Northwest Territories. As with other departments, a sustainable development strategy from DIAND was to be tabled in the House of Commons within two years of Bill C-83 coming into force (fall 1997). The strategy 'should be an integrated approach to economic, social, environmental, and foreign policy. It should establish a framework in which environmental and economic considerations are interlinked and move in the same direction. A single strategy is desirable for the three programs (Corporate Services, Northern Affairs Program, and Indian and Inuit Affairs Program) in DIAND.'[10] A steering committee has been established to make periodic reports to the Senior Policy Committee; consultation with other federal departments and stakeholders will be undertaken. The consultation plan was to be regionally focused.

Negotiations continue between the Inuvialuit and DIAND to overcome an impasse regarding the issuance of prospecting permits and licences. DIAND is 'well-advanced in the work of modernizing the mining royalty regime in the Canada Mining Regulations.'[11] Once a discussion paper is released early in 1996, stakeholders will be able to review the proposed changes. There are no environmental provisions in the Yukon Quartz Mining Act and the Placer Mining Act (which were promulgated in 1924 and 1906 respectively), so DIAND has been preparing a draft bill to amend the mining acts. The intention is to 'provide authority to establish regulatory regimes to place environmental controls on mineral exploration, production, and decommissioning of mine sites. The bill was largely based on the recommendations of the Yukon Mining Advisory Committee that was formed in 1990. Mine-site reclamation regulations will be established soon after the passage of the bill and land-use regulations.'[12] To address another policy, the Northern Mineral Policy was to be updated, and a review paper was being prepared that focused on stakeholders' input; the review was to be based upon the WMI principles and, for the first time, would include sustainable development as a methodology. Among other activities, DIAND

and the government of the Northwest Territories have committed to providing two-thirds of the $11.3 million cost over five years needed to support the West Kitikmeot/Slave Study. The study is a multi-partner initiative that will examine the environmental, socioeconomic issues relating to mineral development in the general area of the Slave geological province. During the week of 20 November 1995, a reorganization of DIAND occurred.

Finally, the Department of Fisheries and Oceans (DFO) has looked at a number of areas related to the WMI, including environmental protection, planning and environmental assessment, and the use of information and science in environmental decision-making. The DFO's immediate policy actions included the following: the Conservation and Protection Guidelines (1994); the Directive on Issuance of [Fisheries Act] Subsection 35(2) Authorizations (finalized, May 1995); and the 1994 Decision Framework for the Determination and Authorization of HADD (Harmful, Alternation, Disruption or Destruction of fish habitat) (1994).[13] The DFO made presentations to the mining industry on the fish habitat management program. Darlene Smith of DFO states that her department's participation in the WMI was important. Officials, for example, could help clarify the requirements of the Fisheries Act to the mining industry. She found the WMI very useful in building understanding among the different stakeholders and thinks that such a process might be useful in other policy sectors.[14]

British Columbia

British Columbia is the first jurisdiction in Canada that has extensively and comprehensively employed 'roundtable' approaches to resource-based decision-making. It is in BC that the diverse interests are most intensively and vocally represented. The province's scenic wilderness and huge undeveloped areas have attracted the attention of many competing interests, ranging from resource developers to tourism operators to deep ecologists. Other countries are now examining British Columbia to determine the weaknesses and strengths of these initiatives and their possible applicability to other situations.

In response to the WMI Accord, British Columbia established the British Columbia Advisory Council on Mining (ACM) in order to build provincially on the national efforts of the mining initiative. In many ways, the provincial follow-up to the national Whitehorse Mining Initiative is an extension of the broader land-use policy approach (the Commission on Resources and Environment – CORE) that was already in 'full swing' in the province. It is important to understand that wider context because successful implementation of the WMI will depend on whether political conditions exist that would encourage long-term support for the integrated vision that it presents. The multi-stakeholder land-use processes under way in British Columbia have already gone through many of the trials and tribulations that the WMI is now facing as it attempts to implement the accord. The CORE Commission was initiated in response to emerging concerns expressed in other forums, including Parks and Wilderness for the '90s, Old Growth Strategy, Dunsmuir Accords, Provincial Round Table, and the Forest Resources Commission. It was recognized that a provincial strategy was needed because of the intensification of conflict over resource use and concerns about the environment. While CORE itself was a flawed process, in part because of its ambitious scope, much was learned from the initiative.

Dan Adamson, a government official with BC Parks who has had considerable experience with roundtable processes, is optimistic about the future and the utility of this approach in land-use decision-making. He suggests that it certainly makes sense from a political perspective: 'When politicians have to make a win-lose decision, they are in a quandary about how to satisfy the broadest interest. If everything goes well, consultation processes develop strategies and solutions which will have a broader level of community support. Government's role is to endorse as long as it fits within government policy and it is determined to have been a fair process.'[15]

Adamson sees public participation in land and resource planning as a positive step in policy-making. He reasons that stakeholder involvement improves the process: 'because we are talking about negotiating values and public servants should not be doing that.' Furthermore, the job of the bureaucrats is not to impute public value.

Consultative processes do not easily resolve troubling questions about how the interests of capital can be integrated into these processes in a way that still provides for an attractive investment climate. On the other hand, Adamson points out that with 'goalposts' established by cabinet in these processes, there is an effort to reach a higher level of land-use certainty than was the case before such planning processes. Under these plans, access to resource use outside of these protected areas is not to be prohibited through these processes.

Nevertheless, the difficulty remains that access may be restricted in a way that could discourage investment. These land- use processes do, however, have a certain allure in that they are more responsive to the values and needs of a wider spectrum of interests. Further, a stable regulatory environment is an essential requirement for capital investment. If the participants of the roundtables themselves can reach agreement, they may be able to create the certainty and harmony in land-use development that sustainable development requires.

British Columbia and the Advisory Council on Mining

Within the context of the province's overall land-use planning process, and building on the principles and recommendations established by the national WMI, the British Columbia government created an Advisory Council on

Mining (ACM) to oversee the implementation of the White-horse Mining Accord on a provincial level. An examination of BC's provincial implementation of the WMI may shed light on the nature of administrative changes brought about in the shift from what former CORE Commissioner Stephen Owen refers to as 'command decision-making' to the 'politics of inclusion.'[16]

On 7 February 1995, Anne Edwards, the minister of Energy, Mines and Petroleum Resources (MEMPR),[17] announced the formation and composition of the Advisory Council on Mining. The minister emphasized the importance of mining as the province's second largest resource industry. The Advisory Council on Mining has the following tasks:

- to develop a stakeholder Statement of Commitment of mining in British Columbia
- to provide the minister with advice on various programs and policy initiatives being considered by the government as a means of implementing the various goals outlined in the WMI Accord
- to provide a multi-stakeholder forum for monitoring and evaluating implementation of the WMI Accord in British Columbia
- preparation of any reports regarding the status of implementation of the WMI Accord in British Columbia that are required for WMI follow-up meetings
- to attend to such other tasks as may be requested or agreed to by the minister from time to time.[18]

For British Columbia, this council fits nicely within the multi-stakeholder, consensus-seeking philosophies discussed above. The government's aim was to ensure that diverse perspectives will be represented. The original list of membership consisted of eleven people and included representatives of labour, environmental groups, mining, Aboriginal peoples, mining communities, and one academic as a member of the public-at-large. Up to fifteen members may be appointed by the government. The council also consisted of the minister (ex officio), the deputy minister, and the assistant deputy minister. In addition, an independent, neutral chairperson, agreed to by all members of the council, would act as a mediator/facilitator to assist the council in achieving its mandate.

It is worth examining the Advisory Council more closely in terms of its implications for the provincial impact of the national WMI. What did the council wish to achieve? How would these goals be implemented? How representative was the council? Would it be successful?

The Mandate of the Advisory Council on Mining

The first few meetings of the Advisory Council were devoted to devising a statement of commitment; that is, a commitment to a healthy, sustainable, and environmentally responsible mining industry (see Appendix C). The challenge was to arrive at a statement about the continuing important role of mining while recognizing that other values must also be protected. Some of the wording of the statement reveals the comprehensive attempts to achieve agreement:

> Our approach is based upon our recognition that the natural environment, the economy, and British Columbia's many cultures and ways of life are complex and fragile, and each is critical to the survival of a modern society. Furthermore, no aspect of social, economic, and environmental sustainability can be pursued in isolation or be the subject of an exclusive focus without detrimentally affecting other aspects. These goals must be pursued in a manner that is flexible enough to accommodate changing economic, environmental, and social requirements.[19]

In spite of the diverse interests that had to be accommodated, the process went relatively smoothly guided by the chair, Dan Johnston, who had steered the national WMI to an agreement. Some of the participants were becoming quite familiar with the roundtable process, having gone through it nationally and by taking part in the many provincial processes of this nature. Members of the council acknowledged each other's right to have their concerns accommodated in the vision statement. While not everyone was necessarily happy with the compromises they had to make in the wording, they agreed to participate in drafting the document. The document was also submitted to a First Nations representative who suggested some wording for the sections pertaining to Aboriginal participation in mining.

The council also addressed issues that affect the future viability of the mineral industry, including the main bottlenecks in the existing mine development assessment process and prioritizing the general impediments to mine development.

In preparation for their two-day September 1995 meeting, discussion papers were presented. Labour representatives responded to this request with some selected issues that they thought the Advisory Council should consider. First, it was suggested that the mining industry should concentrate on working within the 'emerging framework of resource management in B.C. and Canada.' They emphasized the need for partnerships with organized labour and environmental interests. In terms of land-use policy, both the industry and labour have some mutual concerns. Mining activities must be built into regional land-use plans (CORE) as well as the Land and Resource Management Plans (LRMPs) that are being developed at the subregional level. To achieve this, it is necessary to get public support and to provide a public information program 'about the limited environmental impacts of mining, and the broad range of regional and local economic benefits.' Other issues included improving worker skills and retraining; the harmonization of environmental regulations; and some restructuring of the Ministry of Energy, Mines and Petroleum Resources to eliminate potential conflicts between its role as regulator and that of revenue collector. It was also noted that the administration of sub-surface interests below existing Indian Reserves was different in BC than elsewhere in Canada. In BC, there is a split ownership between Canada and the province. A unified arrangement is necessary so that 'mining opportunities involving Aboriginal Settlement lands can be easily explored by both First Nations and others interested in resource extraction.'[20]

An environmental representative suggested a three-pronged strategy. It was first recommended that the BC mining industry assert some global leadership by setting an example to raise standards in other countries where health, safety, and environmental mining practices need improvement. Highland Valley Copper could be used as a 'model mine' of some of the labour and environmental practices possible in today's mining industry. The second recommendation involved various strategies to maintain an ongoing commitment to long-term environmental protection from mining-related impacts. Third, in the area of land-use and environmental assessment, it was emphasized that progress toward meeting the needs of mining must take place in a political environment where land claims, fisheries protection, wilderness tourism, and the protection of biological diversity are also recognized political commitments. Suggestions for meeting the goal of integrating land use with a streamlined environmental assessment process included effective sharing of information by the industry on the state-of-the-art, 'predictive, mitigative and preventative measures'; and the building of expertise on the part of a conservation group with respect to responsible mineral development.[21]

Representatives of the mining industry stressed the need for *'aggressive exploration and mine development.'* It was asserted that, first and foremost, the government needed to be committed to ensuring that mining was considered 'a sustainable and important part of the B.C. economy.' This would include a globally competitive taxation regime to attract investment and timely approvals of projects. Mining representatives also highlighted the importance of clearly defined environmental standards appropriate to the conditions of the project areas. It was argued that environmental performance contracts might, in certain cases, be an alternative to the present system of permits and licences. The settling of Aboriginal land claims with clear ownership would also go a long way toward eliminating uncertainty caused by overlapping decision-making authority. Finally, land-use uncertainty was also identified as a 'major obstacle' to mineral exploration. It was asserted that land-use policy should recognize the unique activities of mining. The industry suggests that 'mining and advanced exploration are so site-specific and likely to occupy such a tiny proportion of the land base, that any conflicts or problems that might arise can be dealt with on a site-specific basis with a process designed for the particular area in question.'[22] Having identified various interests in public consultations, the government should now move toward some 'firm and final land-use decisions,' so that investors know that only protected areas are removed from exploration. This means that 'if you can stake it, explore it, and find a viable mineral deposit and can meet the environmental standards of the day, you can mine it.'[23]

The Ministry of Energy, Mines and Petroleum Resources (MEMPR) presented a list of the changes that it was working on to address some of the concerns raised in an earlier review and the Whitehorse Mining Initiative. The list emphasized the changes the ministry was working toward to improve the climate for mining competitiveness through the Mining Tax Act, the environmental assessment process, permitting, mapping, geoscience, and bringing together First Nations and the industry to communicate perspectives. Particular attention still needs to be given to the issues of mining in the special management areas (as established by the CORE process); developing reasonable access to mineral exploration while maintaining wilderness areas; developing an easily understood and workable regulatory framework; ensuring the regulations regarding proposed contaminated sites are workable for the industry; and ensuring the reclamation security is in place so that reclamation will be undertaken, but not in a manner that renders projects uneconomic.

As the member-at-large, M.L. McAllister suggested that it was time for more attention to be paid to the administration and implementation of the various land-use processes introduced over the last four years. All the stakeholders needed to have a clearer idea about time-lines, accountability structures, and the direction of government policy in terms of resource management and land use in BC. Further, more time was needed for participants to acquire a clearer understanding of the political processes and the administration of provincial resource policy in order to develop some goals that would be politically feasible. In particular, the mining industry would benefit from learning how to further its own goals by working collaboratively with a broader spectrum of interests, including environment, labour, First Nations, etc.

A representative from the First Nations summit also attended the September 1995 meeting and suggested that it would be helpful to develop a direct dialogue between First Nations representatives and mineral stakeholders. The representative emphasized that the question that needs to be explored is: how can we all harmoniously co-exist? It was pointed out that Aboriginal peoples are looking for the same things as anyone else – and that is economic certainty in their communities. The goal is to 'make the economy better for all of us.' Too many people do not understand the treaty process and, therefore, see it as a threat.

Specific attention was paid to the harmonization of assessment processes between the federal and provincial governments. The province's new environmental assessment legislation passed in June of 1995 was supported by business and environmental groups. The Canadian Environmental Assessment Act (CEA) was passed in February 1995. This act has many of the same requirements and characteristics as the BC legislation, yet projects approved under the provincial review process (with full federal involvement) may still have to go through another review under the CEA. In the fall of 1995, the Advisory Council wrote to federal Environment Minister Sheila Copps calling for harmonization of environmental regulations, stressing that it is not in the interests of environmental protection nor sustainable mining to have two sets of uncoordinated rules. Since that time, some progress appears to have been made. Where both jurisdictions have an interest in a project, a cooperative environmental assessment would be undertaken. Cooperation can range from information sharing to joint review panels.[24]

The Advisory Council also requested a workshop to see how mining could fit into the guidelines of special management zones (developed under the CORE process); examine the mine permitting process under the new Environmental Assessment Act; and examine the setup of a direct dialogue between First Nations representatives and mineral stakeholders.

The provincial government reported to the Advisory Council that the government was making progress in terms of balancing the need for mining development with that of protecting wilderness areas. On 21 September 1995, it was announced that the provincial government would designate 30,000 hectares of the Skagit Valley in southern British Columbia as a Class A provincial park and would allow mineral exploration in 2,500 hectares outside the new park boundaries:

> The new provincial park will include and protect key features of the valley such as the rare wild rhododendron flats, the Skagit River, and the spotted owl habitat. The 2,500-hectare area over the valley outside the new park boundary will be available for mineral exploration ... a management plan for this area will be developed to balance resource and environmental needs. Any proposed mining development will be subject to a full environmental assessment review.[25]

The valley is also of historical significance to First Nations peoples in the area and those interests will be acknowledged. The mineral claims in the area are referred to as the Giant Copper property (a copper-gold-silver ore body). Economic benefits from future mine development in the area are thought to be substantial.[26] This initiative took place outside of the ACM, demonstrating that government is attempting to accommodate a variety of diverse public needs and resource values including mineral development. The ACM has yet to take the initiative to develop an agenda that would support government activities of this kind. The ACM has also reviewed four projects under way in the ministry. Each of the projects was undertaken by a multipartite steering committee which will provide reports to the Advisory Council. The projects included the establishment of a skills and training committee (a counterpart organization to the national Mining Industry Training and Adjustment Council), a mine reclamation security policy review, mineral exploration standards and the permitting process, and updating the Mine Health and Safety Code. The report and recommendations of the multi-stakeholder Reclamation Security Policy Task Force Committee were presented to the Advisory Council.

Mine reclamation 'entails the dismantling of buildings and structures, and the stabilization and revitalization of waste rock dumps and tailings ponds (in ore processing, waste particle slurry known as "tailings" is pumped into ponds for storage). A portion of this work occurs during the mine's operating life; however, the majority is usually completed during the mine's operating life.'[27] The mining company pays for the reclamation costs. For some mines, ongoing site management would be necessarily long after closure of the mine. This is the case where acid rock drainage (ARD) occurs. The province wants to have *reasonable assurance* that it will not have to contribute to reclamation costs should, for example, a company go bankrupt with large unsecured reclamation liabilities. The general policy was stated as follows:

> The Province will regulate mine reclamation security in British Columbia to provide reasonable assurance that government funds will not be used to cover reclamation costs. This will be accomplished through a combination of risk management and the posting of security in accordance with a formalized risk assessment process. The Province recognizes that each mine is unique. Accordingly, the Ministry of Energy, Mines and Petroleum Resources will implement its mine reclamation security policies with consideration given to the specific site and company involved.[28]

The criteria for mine reclamation policy emphasized flexibility, transparency, integrity, fairness, incentive, administrative ease, public acceptability, and accountability. Some of the issues raised were related to problems of addressing unknown, infinite liabilities. Within this context, the task force asked what was meant by the provision of 'full security.' Second, industry wanted to be assured an 'exit ticket'; that is, 'a company would be able to surrender to the province the mineral title and all environmental and other liabilities for a property, once all of the conditions of the permit and reclamation plan have been satisfied.'[29] Without that release, the industry states that it cannot provide full financial assurance for reclamation. The Advisory Council could not reach consensus on this issue. It was agreed, however, that upon mine closure, a Certificate of Compliance could be issued that would protect a company from changes to environmental legislation but would not 'lessen their liability with respect to any future contamination.'[30] The industry maintains that all companies should be released from liability once they have satisfied their obligations.

In sum, the task force achieved a consensus in principle on some important issues but was unresolved on others. As such, the Advisory Council on Mining was unwilling to endorse the recommendations until a report was presented that specifically outlined areas on which there was clearly agreement and unresolved areas.

The ACM also reviewed the issue of Special Management Zones (as defined under CORE) and its implications for mining. Under the new regional and sub-regional land-use plans of the province, Crown land has been divided into protected areas, multi-resource zones, special/general/enhanced areas, and agricultural and settlement lands. Protected areas are closed to exploration and mining, but the rest of the area including special management zones (SMZ) is not.[31] The main issue is that under special management zones, 'highly sensitive values' would be accommodated under adaptive resource management approaches. What this actually means in practice for a mining company is unclear. Interestingly enough, although the SMZ appears to be an issue of great concern, only one industry representative showed up to the ministry workshop to consider the implications for the mineral sector. If the industry is concerned about its competitiveness being affected by the policy arena, an important first step is to find out what is being proposed that will affect its ability to explore and develop new properties. Yet few appear to want to devote the time to attend such policy seminars.

By February 1996, the ACM was moving into a position where it could make some recommendations. In addition to suggestions on the four projects undertaken by the industry, it requested that the ministry prepare information regarding its progress in the achievement of the WMI objectives. A provincial election was called for spring and minister Anne Edwards decided not to run for office again. The activity of the ACM was temporarily put on hold. By November 1996, the council had not reconvened.

Representation on the Advisory Council

There are some difficult questions to be resolved in the area of representation in these roundtables. First, as members of the council, what is the role of government representatives? Has their role been reduced to that of stakeholder like the other participants, rather than the public interest in general? In this process, the government's role is not to be that of the neutral interpreter of the public interest, nor is it the mediator of competing interests. That has been given to another individual – a facilitator. In setting up the process in this way, government wishes to ensure that the council may act in as independent a manner as possible. This is a worthwhile goal, yet it raises some worrisome questions about multi-stakeholder advisory processes.

Individuals on the advisory councils themselves are trying to represent a particular perspective, but the range of viewpoints within each of these categories can be very broad. It is difficult for members of the council to represent all the various public perspectives on resource development. A further problem of accurate representation was that the First Nations representatives advised that they would decline to participate in the follow-up WMI process in BC. This decision was made by the Chiefs in Assembly at the First Nations Summit, and it affected all provincial advisory/working committees. Members of the council expressed the sentiment that this was a significant loss; they suggested that, perhaps, individuals from Aboriginal communities would come to the meetings on the understanding that the participation would take place on the basis of observer status. Aboriginal communities are in the process of treaty negotiation. As such, they often prefer to deal with the provincial administration on a government-to-government basis. In addition, their participation in any land-use planning processes must be undertaken with the understanding that any agreements they enter into are carried out on an interim basis pending settlement. Further, mining has not yet been an issue on the agenda of the executive body of the First Nations Summit. The crucial question has been the treaty process itself. This, of course, does not preclude individuals from a particular nation from participating and representing their own particular views on mining as it affects them.

Influence of the Advisory Council

In spite of concerns raised by the representative question, it can be argued (as it has been elsewhere in this book) that policy decisions that benefit from the input of the Advisory Council will receive closer scrutiny by a wide range of perspectives than they would have without the new body. This could have a positive effect on the quality of the ultimate policy. Success in this regard, however, is contingent on the following factors:

- the Advisory Council be able to take some proactive role and is informed enough to be a useful part of the decision-making 'loop.' If the council remains completely outside this process, recommendations may take place in a vacuum, and may not be politically feasible.

- if the advice of the council in aggregate is taken seriously into consideration then the final policy decision may be more reflective of a broader spectrum of concerns than has been the case in the past.
- the minister is able to sell the advice of the ACM to the rest of the cabinet. There are competing interests within government and much has to do with the strength of a particular portfolio and the weight given that ministry within the premier's closest circle of advisors.
- the ACM itself has the ability to reach beyond the mandate of one specific ministry and influence the agendas of other government agencies. This is related to the point above.
- the ACM is able to work together as a team toward certain objectives. This may not be an easy feat, given council members' different world views and the fact that participants were deliberately chosen to represent a diversity of perspectives.
- the interest of the government and the participants can be maintained. The council will have to formulate and work toward some specific goals in order to have a sense of accomplishment and willingness to stay with the process.

The ACM was still at its early stages of development and had just begun to formulate future projects when the meetings ceased. The lack of substantive progress and direction was somewhat frustrating. Members of the council more experienced in roundtable consultations recognized that it takes some time to develop working relationships and to explore different perspectives. Non-industry representatives suggest that the industry in general has yet to internalize the idea that it must now work in cooperation with other stakeholders in the formulation of mining policy. The trend toward public consultation appears to be irreversible and not predicated on the tenure of a particular political administration. As noted in earlier chapters, independent-minded prospectors and miners do not have a predisposition toward collaborative policy approaches.

On the other hand, industry representatives have come a long way, often further than they or anyone else expected, in terms of officially endorsing the principles of the WMI. For their part, industry representatives would like to see other policy communities more supportive of the requirements of mining. The industry recognizes that it needs to participate in the roundtables but is keenly worried that the requirements of mining will be overlooked as the political process attempts to accommodate the diverse perspectives of a multiplicity of other players.

Former mines minister Anne Edwards, who formed the council, stated that the multi-stakeholder advisory groups are valuable because they help to get groups to work together. They are also useful because they hold people to their commitments. She claimed, 'it keeps people honest in what they say – people have to keep their promises; and it helps us work together because talking is a heck of a lot better than being out on the street fighting.'[32] If various stakeholders fail to keep their commitments, others will remind them. Edwards reiterated that, nationally, the WMI was very valuable, and now the ideas must be played out at the provincial level because the provinces hold jurisdiction over natural resources. The national meetings, she said, are very helpful to coordinate the work of the different provinces.

Richard Boyce, of the United Steelworkers of America (and a member of the BC Advisory Council on Mining, as well as a participant in the national WMI), maintains that these processes are 'definitely valuable.' From his experience with the WMI, he has learned that it takes some time for participants to get over the politics of the process and to nudge people from their entrenched positions. He notes: 'We have to continue to participate in this way. We [the Advisory Council] have the potential of achieving something great, if all the players can reach an understanding.' Ed Mankelow, of the BC Wildlife Federation and a member of the Advisory Council, agrees that such councils are very valuable: 'We must sit down and talk together, find out what each other's problems are, and see if there are ways in which we can help each other.' The alternative, Mankelow cautions, is not viable: 'If you work outside the system and set up road blockades, etc., why should you not expect the other side to act destructively as well?' Such roundtables are also helpful, he says, because they provide an opportunity to evaluate one's own position in order to make sure that it is right. 'It is important to check your position and make sure that you have accurate information.' Ed Mankelow articulates the spirit of the Whitehorse Mining Initiative. He is keenly committed to protecting the requirements of the ecosystem and is unwavering in his support of environmental causes. On the other hand, he is also aware of the importance of finding common ground with the mineral industry in order to allow for sustainable mining. Mankelow recognizes that the key to living in harmony with the natural world is through cooperative efforts. His approach is similar to the one expressed by Annie Booth, a professor of environmental ethics and policy, who states: 'Interdependence requires cooperation, and therefore, it requires a voluntary limitation on individual actions and desires in the interest of the greater society.'[33] It is only through this holistic perspective that the complex goals of sustainable development can be achieved.

Although the ACM could play a useful role in reviewing ministry initiatives, it did not undertake independent initiatives. To do this, it needed to have some ideas about what is possible. In their efforts not to dictate the process, government officials took a very low-key approach. As a result of this tactic, the council needed time to feel its way toward an agenda that has meaning and relevance to the diverse participants. Governments, without the advice of advisory councils, are quite able on their own to devise policies that attempt to accommodate a range of public demands. This is clearly illustrated by the four projects they introduced to the Advisory Council on Mining and the Skagit Valley Park decision. Nevertheless, given the mandate of the ACM, there are many initiatives that the council could have pursued. First, in order to move individuals away from their 'stakeholder' position, the council needed to find a broader agenda. This can be found in the work of the national Whitehorse Mining Initiative. Part of the ACM's mandate was to monitor and evaluate the implementation of the WMI Accord in British Columbia. The issue groups in the WMI provided a fairly comprehensive list of desired objectives (see Chapter 5). Those recommendations could serve as a framework for the council and others like it to map out their work. The objectives could be used as a checklist from which the councils could suggest improvements in the formulation, process, and implementation of provincial mining policy.

Second, for it to be relevant, the work of the council needed to go beyond that of the immediate group and the Ministry of Energy, Mines and Petroleum Resources. Moreover, the ACM needed to develop some political clout. Linkages could have been made with other ministries and ministers, including the premier. Other members of the mineral industry, the environmental community, labour, and Aboriginal and mining communities could from time to time be engaged in the activities of the council.

Finally, the council needed to be challenged to pursue some common goals actively. All the members of the ACM needed to work toward some broad objectives. A council can only begin to provide some worthwhile policy direction when it can develop a measure of mutual trust and be in a position to articulate a common position on a variety of issues. Significant progress will be achieved if, and when, the agenda of the ACM is one wherein the participants transcend their own particular interests and develop a world view that accommodates a wider range of perspectives, allowing its members to act in concert.

Unlike the national WMI, the Advisory Council on Mining was not formed with a specific goal to work toward. Also, it was not tied to a deadline. This situation fostered the lack of focus that affected the process. Furthermore, some processes, such as the provincial LRMPs, are now part of the decision-making apparatus. The lines of accountability and reporting relationships are relatively clear. Again, it is unknown whether the advice the council offers will ultimately affect government policy. If the advice is seriously considered and it helps influence decisions, the policy result will be one that has benefitted from a diversity of perspectives. It may, therefore, be more acceptable to members of the interested public. There is, however, a long way for the ACM to go – if it is revived. The signing of the Statement of Commitment, the review of ministry projects, and the letter to the federal minister of environment were only the beginning.

Ontario

In Ontario, land-use planning in the early 1990s had been the subject of many different 'multi-stakeholder' oriented planning task forces and public consultation exercises on land-use reform; the most well known included a commission on Toronto's waterfront (the Crombie Commission) and a Commission on Planning and Development Reform in Ontario. The Commission on Planning and Development Reform, also known as the Sewell Commission, was oriented toward clarifying provincial/municipal roles; streamlining the planning process; having transparent accountable, local decision-making; and adopting strong policies that integrate environmental, economic, and social considerations. When a new Conservative government was elected in 1995, it introduced sweeping radical changes to those and other initiatives and processes. As a result, the future of integrated resource management in the province is currently unpredictable.

Commission on Planning and Development Reform in Ontario

The planning commission was appointed by the minister of Municipal Affairs with a mandate: 'To recommend changes both to the *Planning Act* and to related policy that would restore integrity to the planning process, would make that process more timely and efficient, and would focus more closely on protecting the natural environment.'[34] Unlike the BC Commission on Resources and Environment (CORE) approach, the Commission on Planning and Development Reform had a specific goal of restructuring the Planning Act. Moreover, it was only an advisory body, whereas CORE had a legislative mandate. While many of the activities of CORE seemed to be directed primarily at forest-related issues, the Ontario Sewell Commission focused more on problems associated with urban sprawl. In addition, the work of the Sewell Commission and the Ontario Planning Act applied only to private land (which in Ontario comprises 14 per cent of the land base). In many respects, then, Ontario's and BC's multi-stakeholder approaches to consider land-use planning were fundamentally different. In the case of mining, the recommendations of the Sewell Commission would only affect mineral aggregate extraction and industrial minerals, which are on private land in southern and central Ontario. Crown lands are administered by the Ontario Ministry of Natural Resources, with the exception of mining lands, which are administered by the Ministry of Northern Development and Mines. The majority of mineral exploration and mining of metallic minerals on mining leases or patents occurs in northern Ontario, which would not be directly affected by the recommendations of the planning commission.

Both planning approaches, however, had some elements in common, and both affected mining. Similar to CORE, the Sewell Commission set its key task as finding 'common ground' among the various affected players, including developers, activists, environmentalists, farmers, municipal representatives, and so on. Public participation was encouraged as the commission held dozens of forums and community meetings.

The Sewell Commission released its final report on 21 June 1993. The report emphasized the need for strategic planning and the formulation of a structure that would allow municipalities to address broad planning issues, work jointly with each other, and share scarce resources and limited funds. Public involvement was strongly encouraged. Planning decisions must be consistent with protecting the natural environment and ecosystem while promoting community development. The protection of resource operations was also noted in one of its goals, stated as follows: 'To protect non-renewable resource operations, significant deposits of non-renewable resources (including mineral aggregates, minerals, and petroleum resources), and areas of significant non-renewable resource potential for resource use.'[35] In Northern Ontario, the commission recommended that planning boards be strengthened. Dispute-resolution techniques were to be encouraged in place of having so many of the disputes referred to the Ontario municipal boards. Under the old planning system, all planning authority rested with the province. A new act based on the commission's recommendations allowed for much more regional control, with the province retaining responsibility for overall provincial planning policies, which would address such things as intermunicipal planning issues. The commission also recommended that land-use interests be coordinated by the Ministry of Municipal Affairs across provincial ministries.[36]

Mining and the New Planning Legislation

Mining received some specific attention under the new planning legislation. Under the new Comprehensive Set of Policy Statements (CSPS) that took effect on 28 March 1995, the following policies were presented.

The goals regarding mineral resources, presented under Section F of the CSPS, were somewhat different than those recommended by the Sewell Commission. They were:

> Goal 1: To ensure all parts of Ontario possessing mineral aggregates, an essential non-renewable resource to the overall development of any areas, share a responsibility to identify and protect mineral aggregate resources and legally existing pits and quarries to ensure mineral aggregates are available at a reasonable cost and as close to markets as possible to meet future local, regional, and provincial needs ...
>
> Goal 2: To protect mineral and petroleum resource operations, deposits of minerals and petroleum resources, and areas of potential mineral and petroleum resources for resource use.

> 2.1 Mineral and petroleum resource operations and areas of potential mineral and petroleum resources will be identified for resource use and protected from incompatible development.
> 2.2 In areas of deposits and areas of potential mineral and petroleum resources, development that precludes or hinders future access to and use of these resources will be permitted only if
> a) resource use is not feasible; and
> b) existing or proposed uses serve a greater long-term public interest than does resource use.
> 2.3 Development on lands adjacent to mineral and petroleum resource operations, or adjacent to areas of deposits, will be permitted only if:
> a) the development would not preclude or hinder the continuation of the existing operations; and
> b) the development would not preclude the development of the resource; and
> c) issues of potential public health and safety and environmental protection are addressed.
> 2.4 Rehabilitation of mineral and petroleum resource lands will be required after extraction and other related activities have ceased.[37]

On 9 December 1994, Bill 163, the Planning and Municipal Statute Law Amendment Act, received royal assent. The new act was proclaimed 28 March 1995. In June, a new government under the Progressive Conservative Party was elected, replacing the NDP. They vowed that they would be making some changes to the new planning act. The changes were extensive and involved streamlining its environmental regulatory structure. Those critical of the legislation see the changes as much more than streamlining, voicing concern over the 'loosening [of] key conservation regulations and the laws governing the environmental-assessment process.'[38]

Plans for the WMI in Ontario

After the Whitehorse Mining Initiative Accord, the Ontario government under the NDP moved to establish an advisory council. As with the BC government, the Ontario Ministry of Northern Development and Mines noted that many projects and programs were already under way to implement some of the recommendations of the WMI. Such projects included land-use planning and decision-making processes, mine reclamation, the development of availability of readily understandable mineral resource information. In the area of land-use planning, the Ontario Planning Act described above encapsulates many of the land-use working groups' proposals. As with the proposals of the working group, the new planning act emphasized multi-stakeholder involvement addressing ecological and socioeconomic issues:

- Federal, provincial, or territorial and Aboriginal governments and the minerals industry should work together to ensure that better mineral resource information is more readily available and understandable to land-use decisionmakers. To this end, mineral resource data should be developed and incorporated into land/resource CIS databases.
- A complete Mineral Information Inventory should be done prior to any land-use decision.
- All parties, and especially decisionmakers, who have had meaningful input in a land-use decision-making process, should commit to the implementation and enforcement of the decisions and regulatory controls identified through that process.
- Where candidate protected areas are proposed, or other land-use designations that affect mineral activities, the nature and timing of any constraints imposed on mineral-based activity should be made clear as early on in the decision-making process as possible.
- Land-use planning should strive for clearly defined recommendations that will be implemented and that will provide clarity and certainty on acceptable land/resource uses in each area of land.[39]

The Ontario Ministry of Northern Development and Mines (MNDM) pointed out that it had already been working toward achieving many of the goals outlined by the WMI. The ministry had participated as a plan-review agency with the Ministry of Municipal Affairs (MMA) on the review and approval of official plans and development applications. Planning interests in mineral resources and mine hazards are identified through policies and land-use designations in official plans. Development decisions must conform with the policies in the official plans. Public consultation was built into the approval processes for these documents.

The industry has been concerned about liability issues related to the reclamation of old mine sites, and that a prospector's or company's environmental liabilities are limited to only those they have caused. The Ontario government stated that they have done this as well but that the onus is on the explorationist of old mine sites 'to carefully document any existing damage or disturbances existing before undertaking additional work.'[40]

Former mines minister S. Martel also noted that Ontario had already taken many other initiatives that are in line with the WMI recommendations, as well as the establishment of a Mineral Sector Advisory Council. They included:

> the provision of mineral exploration training for Aboriginal prospectors; the multimillion-dollar expansion and computerization of the province's geoscience database; and the establishment of the Mineral Sector Advisory

> Council to provide the ministry with ongoing advice from the mining industry.
>
> This latter group was to be asked to participate fully as the province moves to implement other recommendations from WMI study groups: Stakeholders from the mining industry and other key groups reviewed recommendations from such areas as education and training, equity financing and investment, investor education, public awareness, and security of title.
>
> The advisory body can make recommendations itself to the minister, or it may decide to undertake further study. It may also set up committees or task forces to examine particular areas of the WMI recommendations.[41]

The Mineral Sector Advisory Council (MSAC) had a few meetings and has identified a number of WMI items for follow-up by government and other stakeholders, including:

(1) Ministry of Northern Development and Mines (MNDM) interact with the Ontario Ministry of Finance and the Ontario Securities Commission (OSC) on matters relating to access to capital and OSC policies
(2) MNDM, Ministry of Natural Resources (MNR), and representatives from other stakeholder groups meet and make specific consensus recommendations for improvements to the land-access process
(3) MNDM convene a group of stakeholders to look at issues relating to appropriate policies for orphaned mine sites and the 'exit-ticket' concept for closed mines. MSAC also endorsed the Minister's Mining Act Advisory Committee and the Ontario Mining Association's position regarding the financial aspects of mine reclamation
(4) Convene an MSAC-sponsored Working Group, under the leadership of the Canadian Aboriginal Minerals Association (CAMA), including First Nations, industry, labour, and government representatives to develop guidelines for aboriginal-industry communication and issue resolution
(5) Encourage members of the Board of the WMI Mining Sector Council to provide regular updates to MSAC on the activities of the council and its various study or working groups. MSAC would provide advice or participate, as may be requested, in the work of the Sector Council.[42]

MNDM had taken a number of actions in the areas emphasized above. Some of the projects were already under way. One of the major issues of concern includes land access for the mineral industry. The deputy minister of the Ministry of Natural Resources stated that the ideas presented in the Whitehorse Mining Initiative would be 'fully considered' in MNR's planning system review and other land-use initiatives such as *Keep It Wild*. It was also pointed out that the WMI process contributed to the development of useful guidelines for mining industry/Aboriginal community clarification.

The Canadian Aboriginal Minerals Association (CAMA) has taken a lead role in gathering input of First Nations concerns with mineral development. It sponsored a multi-stakeholder conference in Sudbury in November 1995 entitled 'Exploring Common Ground – Aboriginal Communities and Base Metal Mining in Canada.'

The intention of the Advisory Council, as the name implies, was to operate at arm's length and to advise the government on various issues related to mining. Participants on the council represent a range of stakeholders and perspectives. It appears that the Advisory Council has yet to seize the initiative to shape its future direction. The council does not meet regularly, something that may be attributed to the fact that it does not have a specific agenda. According to one government official, another multi-stakeholder group has been more active, possibly because it has a specific mandate. The Minister's Mining Act Committee, which was in existence before the WMI, is an advisory group with a specific mandate to focus on changes to the Mining Act. For the Mineral Sector Advisory Council to have an impact, it must become internally cohesive and organized if it is to influence government policy. Furthermore, its members have yet to recognize that they, as stakeholders, need to direct their future structure rather than wait for government to do so. In Ontario's case, as with advisory groups in other provinces, this has yet to happen.

The Mineral Sector Advisory Council would be well positioned to consider and advise issues that are publicly controversial. For example, particular groups are concerned about the potential environmental impact of changes to legislation governing mine-site rehabilitation. The legislation allows for more self-governance on the part of mining companies within certain guidelines. As a multi-stakeholder group, this council could investigate the new regulations and see whether they fit into the spirit of the Whitehorse Mining Initiative and advise various constituencies and the public about their findings. The new regulations are described below.

In October 1995, the new northern development and mines minister, Chris Hodgson, announced changes to the Ontario Mining Act to streamline the mine-site rehabilitation regulatory processes. Currently, all proposed mines must be accompanied by closure plans, which include provision of financial assurance to ensure the rehabilitation of a mine after the mine closes. Such mine plans must receive ministry approval. The intention of the new amendments is to allow the mine-closure plan review process to be more self-regulatory than it had been in the past, through a certified closure plan process. In addition, the financial assurance options have been increased. The ministry will consider a variety of financial assurances that the company will have adequate resources to cover the costs associated with an environmentally acceptable closure plan. The ministry stated that the changes were introduced to allow government to reduce regulatory uncertainty and cut costs, while maintaining its commitment to the environment. Current legislation dealing with the 'polluter pay' principle and the requirement of financial assurances from mining companies in order to cover mine-site rehabilitation costs will be retained. The ministry will also establish 'spot checks' and 'material change' reports from mining companies to ensure compliance. Penalties for failure to comply are $30,000 per day and will be cumulative. Once the mining company has completed its closure plan and decommissioned the mine according to provincial standards, 'the ministry may accept a voluntary surrender of lands in exchange for financial compensation to cover long-term care.'[43] This would release the company from post-decommissioning environmental liabilities. The plans would be filed and available for public viewing. The legislative amendments also deal with public health and safety concerns in relation to abandoned mines and existing mine hazards.

The Canadian Environmental Law Association has stated its concern about these changes and others, claiming that the new government has introduced reforms that will not provide enough environmental protection and safeguards.[44] Another example provides an illustration of these concerns. The Niagara Escarpment Planning and Development Act requires all developments along the landmark to have a permit. Yet a regulation passed in October 1996 exempts long-term quarry operators (those licensed before 10 June 1975) from requiring a permit if they wish to expand their operations. The escarpment is 'the most prominent geological landmark in Southern Ontario.'[45] These are precisely the kinds of issues a multi-stakeholder advisory group could effectively consider. The group would then be in a position to conclude whether or not the principles of the WMI and its Leadership Accord are being upheld in practice. For the provincial advisory bodies to be effective, they have to be publicly recognized as taking the initiative for monitoring and advising on such issues. It appears that neither the BC nor the Ontario advisory groups have yet reached that stage of cohesion or unity of purpose. They will also need government support.

New Brunswick

As was the case with British Columbia and Ontario, New Brunswick also established a land-use commission. The New Brunswick Commission on Land Use and the Rural Environment (CLURE) was set up in January 1992 to recommend land-use policies based on principles of sustainable development. Some of the work to implement the goals of the WMI harmonize with the recommendations of the CLURE report released on 30 March 1993. Also in 1993, the New Brunswick Mineral Resources Policy was released. Its initiatives focused on: mineral ore reserves, value added and diversification, environmental protection, institutional framework, and public information and awareness.[46] New Brunswick echoed the BC and the Ontario governments in asserting that it was already working toward many of the goals that were stated in the WMI Accord. All three of the provinces had previously carried out extensive public consultations through their wide-spreading land-use commissions. It is clear from New Brunswick's progress report that the WMI provided a benchmark point of reference by which to assess the performance of a government.

New Brunswick's Mines and Energy Department initiated the establishment of a working group, with multi-stakeholder representation, to facilitate follow-up and implementation of the WMI in the province. Government meets with industry in New Brunswick two to three times per year and meets with the Prospectors and Developers Association two times a year. Labour representatives meet with the Department of Labour, not Mines and Energy. There are two groups of Aboriginal peoples in New Brunswick that public officials are trying to include in the working group; to date, however, they have not had to meet because there is no mining development in their areas. Thus far, one group has responded to the invitation to join the working group. Government representatives from New Brunswick say that they are making a serious effort to be proactive in the spirit of the WMI. They see such an initiative as important, given, of course, that it is not yet clear where all new mineral deposits will be found; it is seen as better to begin working relationships so that they are well established to deal with future issues.

The assistant deputy minister of Mines and Energy in New Brunswick, Don Barnett, stated that from a national perspective, 'there is a need to get moving ... we've got to start seeing some tangible results [of the WMI]. We achieved a better understanding among stakeholders, and now when it gets down to the task, there is an element of frustration.'[47] Of course, as Barnett astutely pointed out, observers of the WMI must be careful to ask the question: What is the definition of 'fast?' Some governments, he asserted, are pleased with the process and are now saying,'let's get on and proceed.' Our political process, however, takes time to deal with issues, such as reclamation, because one has to measure potential impacts carefully.[48] It is interesting to note that ministers in the WMI consistently agitated for a product from the WMI process, yet now that the task is to implement change, the process seems frustratingly slow to representatives from industry, environment, Aboriginal peoples, and labour. Although the WMI process realized a degree of trust and understanding, conceptual differences remain; Barnett thinks that such differences are healthy. For example, sustainable development from industry's perspective means economic development with environmental protection, and from environment's perspective, it means environmental protection with economic development – the emphasis remains different, as is to be expected. What the WMI achieved, however, is that those two sectors – industry and the environment – as well as the other three sectors (government, Aboriginal peoples, and labour), took time to look at the concepts, in this case sustainable development, from both perspectives. The dialogue and the learning process must not, in Barnett's estimation, be under-valued as an important achievement of the WMI.

There was some concern that once the WMI Accord was signed, people would see it as a moment of closure, rather than as a beginning. The issue groups had identified a number of areas that required continued work, but the tendency might be to 'get on' with business. In Barnett's view, there was a possibility that the stakeholders could fall back into their specific, colloquial views, with an orientation that one issue is important to focus on rather than looking at issues collectively. There were representatives from every stakeholder group who thought that the WMI was the first stage in a long-term process. However, most governments, and industry, too, looked at the process as an expense and pushed for closure on the expensive process. On the other hand, of course, one must consider what the costs of adversarial, combative relations among the stakeholders would have amounted to over the years, without the level of trust that was achieved. Further, the WMI added significant momentum and impetus to pushing governments to streamline the regulatory environment; it was observed by one industry representative that the federal government is now moving on issues of concern to the mining industry and other stakeholders in a way that it would not have without the WMI.

Nova Scotia

Nova Scotia wrote and distributed a new mineral policy for public review and comment; the final report was completed late in 1996. The policy addresses many of the principles and goals cited in the WMI and has been designed to encourage:

- a thorough understanding of the geology and mineral resources of Nova Scotia gained through continuing research and exploration activities
- a business climate that is competitive at national and international levels, and supported by clear, fair, and effective policy and regulatory processes
- greater public support for mineral-based activity and increased public knowledge of the province's geology and mineral resources
- integrated land-use decision-making processes that consider minerals with other resource interests and provide greater certainty and clarity for land access and mineral rights tenure
- protection of the environment through cooperation among regulatory agencies, industry, and the public in supporting environmentally sustainable economic development

- protection of health and safety for workers and the general public through cooperation among regulatory agencies, industry, and labour
- cooperative working relationships between stakeholders, and compatibility with policies, decisions, and actions from other government departments and agencies.[49]

More specifically, a New Environment Act was approved by cabinet in January 1994; a 'one-window' process is under way in which the various government departments involved in mine-development approvals function as a team to simplify the review process for government and the industry; a detailed review of abandoned underground mines is ongoing; an implementation plan for Nova Scotia's Sustainable Development Strategy was developed; Nova Scotia's Planning Act was reviewed; a discussion document was prepared enabling the province to identify resource-management principles and goals and find ways to implement resource planning objectives within the context of municipal planning; and a review of the occupational health and safety mining regulations is continuing. Pat Phelan, the executive director of the Minerals and Energy Branch in Nova Scotia's Department of Natural Resources, states:

> Our department has used the WMI many, many times, most evidently in our new Mineral Policy. The process was useful, we met people, learned how they think about different issues ... the interaction educated people about a lot of things. It also puts your own problems within a national context. It confirmed a lot of things for us as a way to go, such as integrated resource management. All the things going on were influenced by the WMI.[50]

Through all this, the department went through a major reorganization and a degree of downsizing.

Both mining and tourism play a major economic role in a province, as in Nova Scotia. Ensuring that the public has a good understanding of the importance of the mining industry to their economy has been a priority, although Pat Phelan acknowledged that more could be done to communicate the benefits of mining. Don Downe, minister of Natural Resources, emphasized that just because there are some signatures on a piece of paper – the WMI Accord – does not mean that the stakeholders can 'sit back' and watch events unfold. The WMI has been used as a benchmark by which to measure progress in Nova Scotia. Downe states that the department has a very good working relationship with industry; what is needed is communication

> about what is being done with the WMI and to get rid of the image of mining as something that is dirty, takes lives, and bastardizes the environment.

> And to the mining companies we have to communicate the message that the WMI standards are not there to put you out of business but to keep you in it. We have aggressively pursued marketing; there used to be only one marketing person for the whole department, but now we have one designated for mineral promotions. To other provinces, if they have not bought into it [the WMI] to the same degree as we have, then one would have to hope they would see the benefit of making this a true national policy, making this a level playing field. When I sat at the [WMI Leadership Council] table, I was not sitting there solely for the position of Don Downe, minister of Natural Resources, but as a Canadian, as a person wanting to find a resolution, a common ground to move forward.[51]

By the end of 1996, the Nova Scotia government had prepared their new policy document, entitled *Minerals – A Policy for Nova Scotia – 1996*. Seven key objectives of the new mineral policy mirrored the WMI's Leadership Accord. The policy is to be implemented over time. David Hopper of the Department of Natural Resources suggests that 'the WMI provided us with a good foundation to ensure that we are going in the right direction.'

Newfoundland

Newfoundland presented a list of activities that the province noted was in keeping with the spirit of the WMI at the November 1995 WMI implementation meetings. It reported on an improved mining tax regime, which would encourage new mine operations, the reduction of electricity rates, and a reduction of government regulations. Environmental regulations were also analyzed with a view to achieving federal/provincial harmonization (similar to other provincial jurisdictions). A new Mining Act was being developed and would examine issues such as mine reclamation and financial assurance. The province is also working on issues regarding land access, workforce training, and community business opportunities. While the brief document outlined the initiatives it was taking to provide incentives for mining, it did not indicate that significant efforts were being made in terms of fulfilling other goals of the WMI, such as environmental sustainability or providing opportunities for Aboriginal participation in mining.[52]

Since the accord there has not been much implementation in Newfoundland. The WMI consumed enormous amounts of energy from those who participated. Paul Dean, the ADM of Mineral Resources in Newfoundland, observed that some of the momentum of the WMI had dissipated, due in part to fatigue and also to the fact that the leadership council took the glory for the accord; there was no chance for the members of the issue groups to celebrate their accomplishment in achieving consensus, and hence 'there was some reluctance to take up the flag and run again. Further, things got introduced into the leadership council that were not debated around the table in any issue group.'[53] The last point was particularly important for Newfoundland. The leadership council delved into constitutional issues at their July 1994 meeting in Toronto. The Land Access Issue Group had recommended that Native land claims be settled. The leadership council took that idea further to state that Aboriginal peoples have constitutional rights that give them rights to land and resources. Paul Dean suggested that 'some governments will dispute the degree to which constitutional rights extend to lands and resources. That is an interpretation of the Constitution.' Dean concluded: 'We support the process; we thought it was worthwhile. We participated in the Issue Groups and our minister participated in the leadership council. There are ongoing negotiations for settlement of the land claims of the Labrador Inuit and the Labrador Innu; several areas have been withdrawn from further staking until July 1996 and July 1997 so that the negotiations can proceed in good faith.'[54]

Although Newfoundland was not an official signatory to the WMI, there was some progress in issues noted as important objectives in the WMI.[55] In the areas of finance and taxation, the province has amended the tax regime to provide for: (1) the deduction of corporate income tax against mining tax for the first ten years of a new mining operation; (2) rapid depreciation of capital costs; and (3) generous treatment of the processing allowance for mining, milling, and smelting. The provincial corporate income tax rate since 1 January 1995 had been reduced to 14 per cent. New 'economic diversification' legislation was introduced, allowing new mining and other projects to apply for a ten-year holiday from corporate income tax as well as non-profit taxes such as retail sales tax and payroll tax. Interestingly, a regulatory commissioner has been appointed by the province to review all government regulations and to retain only those that are essential for public safety and environmental protection. Environmental assessment legislation was being reviewed and the province is participating in initiatives to achieve harmonization on a national level. Newfoundland was also preparing a new Mining Act, which would address such issues as mine reclamation and financial assurance for reclamation. With respect to land-use policy and planning, a major report on those issues has been approved by the government and is now being prepared for implementation. Various training initiatives have also been introduced, including a basic training program for the Labrador coastal residents to support the potential for new employment opportunities associated with increased mineral exploration; a joint government-union-company retraining initiative for workers at the concentrator of the Iron Ore Company of Canada; and an assessment of the training requirements for the Voisey Bay mining project.

One government official observed that with respect to follow-up on the WMI, 'perhaps the ministers were given a higher profile than they should have ... the perception is that it is up to the ministers to follow up when it is up to *all* stakeholders to do so.'[56] In Newfoundland, much of the exploration activity is undertaken by junior companies; there is considerable frustration by Aboriginal communities in Labrador and Newfoundland with junior companies. Noranda, however, is a company doing what it can to live up to the WMI; the president of Noranda is a signatory to the WMI. The question and the concern is whether junior companies that come from out of province will endorse the WMI and the principles it sets out. The official suggests that: 'Noranda is operating next door to some junior companies from Vancouver who are not doing anything. They have no idea what the WMI is.'

This issue in Labrador raises an important question. Whose responsibility is it to inform junior companies and to encourage, if not to require, them to live up to the WMI? In its status report presented at the 23 November 1995 meeting, the Inuit Tapirisat of Canada state:

> The mining industry, particularly the junior exploration companies need to communicate directly with the communities around which they are staking. NRCan [Natural Resources Canada] and DIAND [Department of Indian Affairs and Northern Development] should be making these companies aware of the land-access and permitting requirements that exist under the land claims, so a situation like in the Inuvialuit Settlement area is not repeated. Nunavut has put out advertising in various papers that outlines the permitting requirements for exploration in Nunavut.[57]

The Inuit Tapirisat place the onus of responsibility for informing junior companies on government; however, industry associations are obliged to share that responsibility. Most junior companies are not members of the MAC but are members of the PDAC; hence, responsibility to educate its membership likely falls to PDAC. One idea, proposed at the 23 November 1995 reunion of WMI stakeholders, was to give a copy of the WMI to new mining association members and, perhaps, to require them to sign a statement of principle that they would agree to implement the accord. In the end, the benefits of acting on the recommendations of the WMI may be the best incentive; 'once companies see people who are acting on the WMI getting along better with Aboriginal groups' they may follow suit.[58] However, 'to be fair to junior companies, they are working with a very small budget and a much smaller time-frame [to produce results for their shareholders]. The playing field [between juniors and a large company like Noranda] is not even. Most large companies really feel this is a modern vision of the mining industry. The MAC is an umbrella for them. What is needed are case histories about the products of the WMI. We need reports about success stories.'[59] Hence, just as good communication and information flows were crucial during the WMI process leading up to the accord, so are such links crucial in the follow-up stage. It is crucial to have some reporting mechanism that allows those in Newfoundland and other areas to hear the voluntary collaborative actions being taken by smaller companies in Saskatchewan and Manitoba. One cannot rely on the media to communicate such success stories. Follow-up to the WMI includes ensuring the establishment of effective communication links; the onus likely falls to industry associations to report the initiatives of their members, but there should be national, not just provincial or regional, reporting links.

Saskatchewan

Saskatchewan had a series of parallel processes that examined the state of the mining industry in the province. A Saskatchewan Mining Task Force was set up in November 1992 'to establish a process to address a number of issues affecting the future development of the mining industry in Saskatchewan.'[60] Its members included five industry representatives and Saskatchewan Energy and Mines (SEM) staff. An issues paper was developed that looked at the impediments to growth and development of the mining industry in the province. The paper was submitted for consideration by government and industry. Eventually, seven other provincial government departments were represented on the task force, which meant that the range of issues it considered was also expanded. The Saskatchewan Mining Task Force was tabled in September 1994 with the minister of SEM. The issues addressed in the report included some of those that the WMI considered: land access and security of tenure; the industry's competitive position; environmental issues; and regulatory changes and requirements. The minister accepted the task force report and agreed to participate in the WMI. The report, states the ADM of Energy and Mines, Dan McFadyen, 'had a lot of correspondence with a number of things in the WMI, but it was not the same thing ... When the WMI came up, our task force was a useful way of keeping our sister departments informed of the WMI's progress.' That is not something to be underestimated. In some provinces, it was challenging to keep officials throughout the natural resources department informed about the WMI, let alone other departments such as Environment or Finance. The task force was a good communications vehicle in the Saskatchewan government.

The task force was only one of a number of processes established by the Saskatchewan government. Others included:

- Partnership for Renewal is a long-term development strategy for the Saskatchewan economy.
- Mining is one of the six strategic clusters identified by the strategy

that generates provincial wealth and employment opportunities. The province must ensure that the industry has the ability to compete in the international marketplace.
- Saskatchewan's Environmental Agenda: Securing a Sustainable Future is a draft integrated policy framework outlining the government's direction for the environment and sustainable development, where securing the environmental and economic future of our communities requires striking a balance between environmental protection and economic growth.
- Mining is identified as an industry that must be committed to the goal of sustainable development, where securing the environmental and economic future of our communities requires striking a balance between environmental protection and economic growth. The mining industry currently disturbs only a small land base, but it can have a long-term impact on air and water resources. Effective operating and decommissioning of mines are essential.
- The Government's Position on Proposed Uranium Mining Developments in Northern Saskatchewan is the provincial government's response to the report of the joint federal/provincial environmental assessment panel on proposals for three uranium mining developments in northern Saskatchewan.[61]

The report reiterated the government's goal of supporting the expansion of the mining industry based on the principles of sustainable development. Saskatchewan put parallel decision-making and consultative processes in place, which resulted in a commitment by the government, across departmental lines, to concepts such as sustainable development. Such a comprehensive approach was required, given that, as elsewhere in the country, the industry was facing declining revenues, poor levels of profitability, mine closures, and job losses. In a province dependent on its natural resources, the government was obliged to tackle those challenges in a serious way. New challenges, such as treaty land entitlements, Metis land claims, and designations for parks and wilderness areas required a comprehensive collaborative process rather than a piecemeal approach to policy-making.

The WMI was an important part of the parallel policy approach that Saskatchewan had established. McFadyen states that the WMI 'as a process was useful. It was the first time such diverse interests rolled up their shirtsleeves and talked in a less confrontational way. They weren't antagonists ... there was a sense of common undertaking. We had a lot of fruitful discussions.'[62] He observes that:

> governments' agendas in the past few years have been driven by the notion that consensus is a healthy objective to strive for ... As a process, the WMI was successful in achieving broad general consensus. One of the

challenges is for government to retain its role as one of five stakeholder groups. There is a lot of pressure from all parties that governments were the owners of the process; that is, that governments were the owners of the product. It was not government that initiated the process; all parties accepted ownership.'[63]

From his perspective, the WMI established a set of fundamental beliefs that will always be at the forefront of discussions.

The minister of Saskatchewan Energy and Mines stated that he was not opposed to people who sought to establish ecological reserves,[64] but rather that he is in favour of setting up environmentally protected networks. After the task force and the WMI, however, the minister emphasized that there are decision-making processes that must be followed. Some people expected that the WMI was 'the answer.' Consistently, most participants in the WMI state that the most important result of the WMI was the design of a process for adding the complex web of issues affecting the mining sector. Another government official concurred, stating that the WMI 'was not the answer ... but the start to a solution. Other jurisdictions do not have exactly the same views [on the issues addressed by the WMI] as everyone else, so implementation has to be on a local basis. We're at the point where we have to stop talking and start implementing. We have these broad principles and we need to move on ... WMI defined the mountain. At some point we have to get down off the mountain and solve these things.'[65]

Saskatchewan has been making progress in some of these areas. In response to the needs of Aboriginal peoples as related to the mining sector, for example, efforts have been made to 'increase the level of Aboriginal participation in the workforce; facilitate the Aboriginal ownership of Crown mineral and surface rights under the Treaty Land Entitlement Framework Agreement; increase knowledge in the Aboriginal community of non-renewable resources.'[66] The government 'has a long history of fairly proactive relations with First Nations communities, particularly in the mining sector. There was nothing different on the leadership council table that we would not have expected to hear here in the province.'[67] Saskatchewan is building relations and furthering tripartite programs between the industry, government, and Aboriginal peoples. The focus of the efforts has been to establish, in concert with other departments, employment equity hiring, a mentor program, cross-cultural training, identification of relevant materials to be translated into Indian dialects, and identification of Indian and Metis training opportunities. On the same date that the WMI was signed, a series of communities and companies agreed to increase Aboriginal employment in the mining sector. The Native Labour Market Committee is an initiative to address Native training and education, upholding the principles expressed in the WMI. Working relations and

Natives' fairly high participation rate have some other provinces taking notice of Saskatchewan's achievement. Nevertheless, it is the political will to tackle the problem of employment among First Nations peoples in a collaborative way that distinguishes Saskatchewan.

One of the dilemmas of the mining industry in Saskatchewan is that the WMI was 'by and large fostered by the Mining Association of Canada.' As noted in Chapter 5, none of Saskatchewan's major mining companies are members of MAC, even though they have a large mineral sector. The assistant deputy minister pointed out that: 'We had representation on the WMI from the uranium and potash sector, but they have different associations and share a common set of values' that might be different than those shared by members of the MAC; 'MAC represents metal miners and the PDAC represents mostly junior companies. We don't have a large metal-mining sector. Each mining industry has a different set of issues to deal with. This, in part, was one of the difficulties of the WMI – it is unfortunate that the MAC got labelled nationally with representing the mining industry. It does represent large and important organizations, but it doesn't represent as a whole, the mining industry; that was their dilemma.'[68] Within the mining industry, then, one does find different levels of commitment to the principles of the WMI. The Canadian Mineral Industry Federation was created since the WMI was initiated in 1992. This federation could serve as an umbrella organization for the industry.

Manitoba

In Manitoba, as in Saskatchewan, progress was being made in terms of collaborative decision-making. This would have been more difficult, if not impossible, to achieve without the WMI. Ed Huebert, executive vice-president of the Manitoba Mining Association, states that there are three efforts in Manitoba that reveal, in a symbolic way, that the way the mining industry is doing business with other stakeholders would not have happened before the WMI process.

First, the Grass River Project was initiated in October 1994. That area is environmentally important, with its very scenic white-water river and one of the main watersheds in northern Manitoba, and it is the heartland of Manitoba mining. In October 1994, the area was being promoted for the purposes of eco-tourism and efforts were made to have the area designated as a Heritage River. The Manitoba prospectors and developers were very concerned about the impact such a designation would have on their activities. Such designations left all stakeholders with strong perspectives and feelings. Other stakeholders included trappers, owners of lodges, eco-tourism promoters, and community residents themselves. After distribution of the WMI Accord, at a meeting by a representative of the Manitoba Mining Association, the members of the association decided to meet with people who had a stake in changes to the region to determine what interests they all had in common. A working group was established to see how eco-tourism and its criteria could be respected while ensuring that mining and exploration could take place in the region. Land use in the region would be examined from a variety of perspectives.

Ed Huebert, vice-president of the Mining Association of Manitoba, stated: 'Rather than having any interest compete for land designation, everyone came together in a spirit of cooperation. The approach followed the principles of the WMI.'[69] The challenge was to think about the 'community and economic base of a region, to consider multi-use land use, to avoid backing a single user – mining shouldn't take precedent over eco-tourism, and vice versa.'[70] The challenge was met. The group, which had a permanent project manager, was successful. Some individuals also looked at what could be done with closed mine sites as a part of developing eco-tourism. But it was not easy to reach consensus. Ed Huebert remembers: 'It took two meetings and a lot of faxes to get people to focus on a common process that recognized various regional perspectives. The first condition was that all local stakeholders must be involved ... But before people communicated their first line of position, all local residents put their position forward, and the provincial government was able to observe all this. One of the first agreements was that we didn't want to advertise this until the project manager was comfortable that a consensus on process was achieved.'[71]

Members of the Mining Association of Manitoba were aware of the WMI, and initially there were a lot of questions and concerns about what the WMI had been about. In Manitoba there is a certain degree of a multi-stakeholder orientation, which may be reflective of the political culture, more so than in other jurisdictions. This may be due to the relative importance of mining to the Manitoba economy, as a proportion of the Gross Provincial Product. The public may, therefore, be more aware of the importance of mining, although archaic views likely exist to some extent. Further, Premier Filmon ensured that his Roundtable on the Economy and the Environment, established in 1989, included representatives from the mining industry. Both the minister and the director of mines were supportive of getting people to talk about the need to achieve access and dialogue with stakeholders, rather than staying in a win-lose orientation. Manitoba, like Saskatchewan, saw the need to decrease uncertainty for the industry, giving it a strong motivation to participate in exercises such as the WMI.

A second initiative reflects the strong success in Manitoba in moving toward cooperative multi-stakeholder initiatives to balance and meet economic and environmental concerns. Some discussions took place between the Mining Association of Manitoba (MAM) and regional representatives of the World Wildlife Fund (WWF). Communications were established to dispel perceptions that mining is incompatible with protected areas. Ed

Huebert said that it 'was the WMI document which had us start to talk.'[72] At an endangered species meeting attended by the MAM, an invitation was extended to Gail Whelan-Enns of the WWF to talk to the MAM's Exploration Committee. The Mining Association of Manitoba understood that there was a social need to protect some areas. Discussions continued between the WWF and the MAM. From the WMI, it is clear that personal relationships are critical to achieving consensus. Huebert states: 'You have to establish trust. It is a relationship-building process.'[73] He observes that 'government people have been wondering what we're doing working with the WWF ... they thought we were on opposite sides of the fence!'[74] And now those discussions have initiated a process that has to unfold through deed.

In addition, the association has been building other relationships, with Aboriginal peoples and with government departments such as Finance. Hence, the WMI was instrumental in motivating different sectors to collaborate in tackling issues identified in the accord. In preparation for a national meeting on the WMI to discuss progress in working toward the goals of the initiative, Manitoba's Department of Energy and Mineral Mines (E&M) went to all stakeholders to collect their views about progress. One official emphasized that Manitoba is the 'first province to embrace the concept of sustainable development. We rewrote the Mines Act and replaced it with the Mines and Minerals Act in April 1992. In that Act, we incorporated eleven principles of sustainable development.'[75] In its submission to the 23 November 1995 meeting of WMI stakeholders (referred to as The Green Report by participants), Manitoba's Energy and Mines progress report stated:

> Many economic development and environment protection issues currently end in an adversarial 'either/or' situation. Solutions put forward do not always effectively resolve the environment/development dilemma. However, a policy framework and process that is based upon the principles of sustainable development provides a different option. This concept calls for a new era of economic growth in which policies for development are based upon factors that sustain environment and at the same time promote the need for economic development.[76]

Manitoba Energy and Mines made an effort to include other ministries in the WMI. Art Ball observes that one of the problems for government with a process like the WMI was that mining ministers and their civil servants were involved, but the policy sector also involved issues that required the involvement or at least, the agreement, of other ministries. An invitation, for example, was extended to Manitoba Environment and was accepted. It was then vetoed on the basis that such participation was not affordable. Manitoba Energy and Mines still kept their counterparts apprised about the WMI. In another province, a few stakeholders have observed that they were envious of Manitoba: 'the government there is very supportive ... the premier comes to their mining events.'

The above overview of some of the government initiatives in the one year following the WMI Accord serves to illustrate just a few ways in which governments adopted some of the goals of the WMI. The WMI has been most successful in those jurisdictions that had already embarked on other multi-stakeholder processes and, as such, had a political culture that was more receptive to the goals of the initiative. For those mines ministries actively promoting the spirit of the accord, much has been achieved in terms of developing cooperative relationships with other ministries and diverse public stakeholder groups.

Nevertheless, the spirit of the WMI may take a long time to percolate through to the members of the different mining and other organizations. Individuals within certain stakeholder groups that did not attend the WMI will have to be sold on the goals of the process. This is certainly the case with many junior mining companies. Others, such as members of the environmental community and Aboriginal representatives, have yet to be persuaded that the industry in general is committed to the initiative, especially when they observe the disinterest on the part of general members of the PDAC or other members of the mineral industry who did not buy into the process. Furthermore, not all governments, nor all ministries, wholeheartedly adopted the recommendations in a way that would see them meaningfully implemented. The period from September 1992 (the signing of the WMI Accord) to November 1995 (the presentation of progress reports) is generally referred to as Phase One of the accord. Many of the proponents of the accord still have a big job ahead of them in selling the vision of the WMI to their colleagues, their constituents, and the general public. The concluding chapters consider the relative success of the WMI Accord and whether such a process can prove to be a useful mechanism for resolving conflict over resource-based decision-making.

7
Perspectives on the Accord

The WMI and Its Related Initiatives

At the end of the day, what has been the contribution of the Whitehorse Mining Initiative? Much depends, of course, on the criteria used to evaluate success. The initial goal of the WMI was to improve the competitive position of the mineral industry by fostering a more supportive policy environment for mining. Such an environment requires the support of the governing administration as well as effective instruments for carrying out the wishes of the elected representatives.

To achieve the political will to ensure a favourable policy environment, there has to be public support. The nature of public support is difficult to assess because it is mediated and shaped both by the media and interest group leaders. It is reasonably safe to assume, however, that the urban majority (particularly in Canada's larger cities) tends to be relatively indifferent to the mineral industry. Opinion leaders, however, can play an important role in influencing public opinion or by influencing public agendas. In this respect, some participants have observed that the Whitehorse Mining Initiative has served as a useful vehicle for bringing a wider constituency into the consultation process. When a broader set of views is used to help influence the policy process, the ultimate decision is more likely to be generally accepted and understood.

In addition, the WMI has served some useful political purposes because it has received a fair amount of positive attention within other government ministries. It has also helped coordinate and set a framework for guiding future mineral policies in Canada. That said, members of various stakeholder groups have expressed reservations about what they see as a weak commitment on the part of industry and government in implementing the goals of the WMI. Much work needs to be done if the goals and objectives of the WMI are to go beyond the spoken commitments of those who participated in the initiative and to be adopted by grassroots members of the various organizations.

What Is Success?

The creators of the WMI were not at all sure what it was that they were unleashing, but they knew the time had come for the mining industry to learn how to work in a more compatible way with other communities of interest. It is important to note that the vast majority of those participants interviewed, which represents a good cross-section of the diverse participants, thought that the WMI was successful on many grounds. The responses ranged from wholehearted endorsement to the more tepid comment that the WMI was probably a useful exercise. Participants frequently commented that they were very surprised that the WMI was able to achieve so much.

Success was defined in a multitude of different ways. For some, it was enough that the final accord was signed by most of the representatives and governments. Many pointed out how much they learned about many different ways in which natural resources could be valued. Learning new terminology that helped explain the differing world views became an important part of that exercise. Some found that they could respect the positions of some of the participants – people who were once viewed as enemies. Individuals often surprised themselves about how far they were willing to compromise. Others pointed out the extent to which people shifted from earlier entrenched positions to positions that were more conciliatory in spirit. Many emphasized the importance of the initiative for developing national contacts between different stakeholder groups as well as within their own sector.

From a public administration point of view, the Whitehorse Mining Initiative provided a framework for interdepartmental discussions and cooperation. Public servants in the mines ministries and departments could point out to their colleagues that they had support from many of the client groups of the different ministries. This has facilitated interdepartmental support and sympathy for the goals of the WMI. It has also affirmed the importance of the industry to the Canadian economy.

Participants generally pointed to four factors when asked what led to the successful signing of the accord: visionary leadership, expert guidance, financial and human resources, and a desire to succeed. Two individuals stood out during the whole process, commanding the respect of participants from all the stakeholder groups: George Miller (president of MAC) and Dan Johnston (the key facilitator). While initiatives of this kind rely on the skills of many people, there are always a few individuals who show leadership in a way that makes others believe that something is both possible and desirable. George Miller was able to champion the WMI at crucial times, encouraging people to forge on when agreement seemed impossible – or at least highly improbable! Miller provided the visionary leadership; this was particularly important because it came from an industry representative. On the other hand, Dan Johnston offered expert guidance. Johnston was a facilitator who had seen his way through many such processes, particularly in British Columbia's fractious land-use policy environment. He was able to gain the trust of the participants by dealing with them fairly and by consistently making sure that they could all contribute on a level playing field. Johnston also made sure that the participants worked steadily toward a goal.

The process was fairly well funded. Financial and human resources were provided to allow a broad cross-section of policy communities to participate. The legitimacy of the accord would have been weakened without such diverse representation. The desire to succeed was also very strong. When individuals put a great deal of time and effort into an initiative and that work gains momentum, people want very much to achieve something positive, such as the signing of the accord. Moreover, many participants were acutely aware of the need to succeed. The conflict that now surrounds most new resource-development projects necessitates a new approach to reduce the level of dissension. People either had to sit down around a table and come to some common understanding or once more face decision-making paralysis as conflicting groups battle over what constitutes the appropriate use of a resource.

By September 1994, the participants were ready to endorse the accord and present their formal perspective of the WMI (presented under the headings below). Many were very encouraged at that time. A year later, however, when implementation was not what some had hoped, the positive responses of many of the NGOs and other stakeholders became more subdued. The following is an overview of the positions of the different policy communities upon the signing of the accord.

The Aboriginal Perspective on the Whitehorse Mining Initiative

> Mineral exploration and development has occurred in many parts of Canada on lands where Aboriginal peoples have maintained a use, affinity for the land, and occupation for thousands of years. Aboriginal communities have depended on the land for a warehouse of natural resources for survival. In the past, mineral activity has disrupted the traditional lifestyle and has left few traditional economic opportunities for Aboriginal peoples.
>
> Aboriginal peoples have not always been consulted on mineral activity nor invited to participate in environmental and infrastructure planning related to a mine project on their lands. They have not received cash compensation, been included in business opportunities, nor been offered opportunities for quality training and quality employment. Similarly, environmental organizations do not consult Aboriginal peoples. In many cases, environmentalists' lack of respect for community economic-development

> goals and objectives hinder community well-being and economic growth. On the other hand, poor environmental management practices and lack of a holistic management approach, often leaves the landscape uninhabitable for Aboriginal peoples when a mine closes.
>
> Recommendations to bind and improve relations between Aboriginal peoples and the mining industry, for present and future generations, will come out of the Whitehorse Mining Initiative.[1]

It is difficult to make general assertions about the conclusions of any of the stakeholder groups because they are so internally diverse. This is particularly the case with the representation of Aboriginal peoples of Canada who came from distinct and separate nations and cultures. It is fair to say that given the long history of Aboriginal peoples' interaction with resource developments over the past centuries, there remains a degree of scepticism that will not be eliminated by the signing of one accord. Furthermore, the Assembly of First Nations felt they could not sign the document because there had not yet been widespread consultations with Aboriginal communities throughout the country. Moreover, bands and treaty groups would also have to be consulted about the possible impact of the WMI on their interests.

Aboriginal people did participate in the WMI process and were active in consulting with Native communities. While the spirit of the accord was endorsed, there are many reservations about its implementation. Jerry Asp, who participated on behalf of the Aboriginal sector in the Land Access Issue Group, stated that the WMI itself was successful because it was able to bring stakeholders together and produce a document that had consensus. However, whether or not the results are successful, Asp states, depends on whether a project can put the WMI itself to the test using all the principles, with equal weight given to each one. Hans Matthews, president of the Canadian Aboriginal Minerals Association, observes, 'The task now before us is for everyone to "rerun" the whole WMI philosophy at the community level with individual communities. Even then, individual communities may have different principles, goals, and recommendations. Some communities will clearly not support the mining industry. For those who do, can we ensure that the spirit and intent of the WMI benefits all parties involved in mineral projects before metal prices take another dive?'[2] In terms of impediments to success, Asp emphasizes something that has been stated by many participants – that the junior exploration companies were not part of the process and might not buy into it. From the Aboriginal peoples' perspective, Asp was not satisfied with the degree of implementation of the WMI Accord one year after its signing: 'From all the progress reports that I have read, it seems to me that the Aboriginal issues are being left on the back burner. I feel things, such as 12 per cent of the land mass will be parks, before Natives are even considered ...' In sum, Asp concluded that

'The work produced by the WMI is good. But without an implementation strategy, it is just another "good" report. I can see implementation as an educational exercise as opposed to [being] forced.'[3]

Other representatives of the Aboriginal peoples of Canada were also active in the WMI. The National Indian Brotherhood, Assembly of First Nations, for example, conducted an extensive survey questionnaire, asking its members to respond to the various topics addressed by the WMI issue groups. Representatives of the Congress of Aboriginal Peoples and the Inuit Tapirisat of Canada (ITC) did sign the accord. The ITC made the following policy suggestions:

> The goals and objectives of the WMI should be incorporated in any reviews and amendments to policy whether national or provincial. NRCan has a responsibility to make mining companies, especially the junior exploration companies aware of the WMI. Many of these companies do not know about it.
>
> There is also a need for governments to recognize provisions of the land claims, like 'Article 24 of the Nunavut Agreement,' which gives Inuit-owned companies priority in the procurement of service contracts. It is not good enough for mining companies to hire a percentage of Aboriginal employees, but rather the hiring and service contracts should be handled by an Inuit-owned company itself. An Inuit-owned company will be better able to assess the needs of Inuit workers than a mining company that usually deals with flying up workers from the south.[4]

The ITC also undertook to fulfil the goals of the WMI on a community-by-community basis in the Arctic. They were involved in environmental protection, environmental assessment and monitoring, protected areas, decision-making, land-use planning, and negotiating Impact Benefit Agreements to ensure that communities benefitted economically from mining development. The ITC will continue to participate in the national WMI Advisory Council.

Industry Perspective on the Whitehorse Mining Initiative

> The many issues facing Canada's mining industry are both domestic and international in origin. Beyond Canada's borders, global factors will continue to be an important influence on the fortunes of Canada's mining industry. These are outside our control. However, we can and must address those factors within Canada which shape the domestic policy and regulatory environment, and which have a major impact on the competitiveness of Canada's mining industry. Making the necessary improvements to the

> complex and dynamic policy and regulatory scene in Canada requires the efforts of everyone. It was for this reason that industry suggested a broadly-based consultation with key stakeholder groups. The Whitehorse Mining Initiative's outputs will be both tangible and non-tangible. Principles, goals, and recommendations will be directed at each stakeholder group, and at policy and regulation. Equally important, the Whitehorse Mining Initiative can open avenues of dialogue, increase understanding amongst the respective stakeholders, and facilitate behavioural changes on all sides. A successful conclusion of the Whitehorse Mining Initiative will help to assure the future of mining in Canada.[5]

The various industry organizations, such as the Mining Association of Canada (MAC), the Ontario Mining Association, and the Prospectors and Developers Association of Canada (PDAC), recognized that they would have to actively inform and involve other industry people who did not directly participate in the WMI consultations. The Mining Association of Canada freely distributed the accord and the issue group reports, making sure that they were widely available. The associations also engaged in various public information sessions.

The MAC also updated its environmental policy (the original, created in 1989, was the first of its kind in the world). The MAC Environmental Committee is proposing to hold a workshop on encouraging community involvement in the mining industry. The PDAC has concentrated on improving the investment climate for exploration and on improving processes associated with land use decision-making. They are also attempting to establish better communications between the mineral industry and Aboriginal peoples.

Industry is working jointly with labour on the creation of the Mining Industry Training and Adjustment Council (MITAC). MITAC is a national initiative to enhance 'worker health and safety, productivity, mobility, and adjustment.' It will also be used to encourage Aboriginal participation in mining. Under the BC Advisory Council, a skills and training initiative is a provincial effort related to the goals of MITAC. The MAC and PDAC have also participated in the minister of natural resources' national advisory committee on WMI implementation.

Some industry association leaders are actively promoting the WMI. For industry commitment to be sustained, however, the grassroots members of the organizations must also be supportive of the WMI goals. This is a challenge since they have not benefitted from direct participation in the extensive consultation efforts.

The managing director of the PDAC, Tony Andrews, stated that the process was a success from his own point of view and that of the organization:

> It identified clearly and concisely the major impediments to our industry, and that was extremely useful to governments because they were able to use the results as a vehicle to make progress on issues and to introduce new initiatives. To a significant degree it brought the industry sector – including government – together ... It created allies. It clearly demonstrated that industry and Aboriginal peoples could be allies. Aboriginal peoples and the industry are not really far apart on many key issues. WMI is only a beginning.[6]

Andrews cautioned that there are many people in the industry who were not involved in the process and do not think that it was a success. He says that the biggest challenge the industry faces is to get people to buy into the vision of the WMI. In summary, he concludes that the WMI was successful because it raised a level of awareness around all the issues, even 'if it was a first chip at the base of the mountain.'

One of the steps up the mountain since the accord has been the establishment of a joint PDAC/MAC lands committee. The committee formed a natural task force to develop a land access strategy based on the principles of the WMI. The strategy has been accepted by the Canadian Mineral Industry Federation (an alliance of industry executives), and it is based on the principles of conservation of biodiversity and protected areas. Outside of the protected areas, the industry supports 100 per cent access.

Some of the participants were not looking for too much from the initiative. Walter Segsworth of Westmin Resources stated simply: 'It was a good process. It worked. It got a result.'[7] The long-term test, of course, will be whether the activities of industry and other participants will be consistent with the WMI principles and goals.

The Canadian Institute of Mining, Metallurgy and Petroleum (CIM) Perspective on the Whitehorse Mining Initiative

> Founded in 1898, the Canadian Institute of Mining, Metallurgy and Petroleum (CIM), is the sole national technical organization in the minerals and petroleum industry in Canada. CIM represents twenty thousand (20,000) people in its dozens of Branches, Sections, Divisions and Societies, from Victoria to St. John's, from Toronto to Baffin Island. CIM sought out involvement in the Whitehorse Mining Initiative (WMI) because it firmly believed in the process and its prospects.
>
> For our members, living as they do in the communities that produce Canada's mineral wealth, the mining industry puts bread and butter on their tables. Its impact is not an abstract concept; its importance is not hypothetical. It is real.
>
> Our members are sensitive to the wilderness – because they live near it

> and work in it. They are aware of the importance of the environment; as scientists and engineers, it is they who will help to ensure that sustainable economical development is reality.
>
> They are conscious of the need for solutions now – because they believe in the future.
>
> For the members of CIM, the WMI brings people together, that are, like CIM, a reflection of Canada itself. We support the WMI and we look forward to its results being representative of a process of genuine, heartfelt collective wisdom. We believe that the process represented by this accord is the beginning of a new era in the mineral industry – one that will secure our country's economic future.[8]

The Canadian Institute of Mining, Metallurgy and Petroleum has also actively engaged in publicizing the efforts of the WMI and promoted the accord. It too adopted an environmental statement. Its strategic plan is to assist and promote mining that reflects the spirit of the WMI.

The Labour Perspective on the Whitehorse Mining Initiative

> Three features of the Whitehorse Mining Initiative made participation attractive for the labour movement. First, through its involvement of the mining industry as a whole, it offered the opportunity to bring forward issues directly affecting our membership and their communities that go beyond those traditionally dealt with in collective bargaining, such as training, work reorganization, closure planning, and long-distance commuting operations. Second, through the involvement of other stakeholders – the government, Aboriginal peoples, and the environmental community – it offered the opportunity to integrate the union's work in other areas with its core collective bargaining responsibilities. Third, the union felt that its experience in conflict resolution and its independent knowledge of the mining industry could make a positive contribution to the process of consensus-building on a multijurisdictional basis contemplated by the Initiative.[9]

Throughout the WMI, labour representatives focused their resources on the Workplace/Workforce/Community Issue Group. Labour was primarily concerned with such issues as work reorganization and technological change, skills training and upgrading, better design of the workplace, etc. A representative from the United Steelworkers of America (USWA) strongly endorsed the creation of a 'sector council' for the mining industry. Further, the suggestion was made that a structure be developed that provided for national standards, with a national apprenticeship committee that would be coordinated with provincial activities. Other suggestions included joint ventures or partnerships 'whether it is simply a new piece of technology, work schedules, the design of how we work, and training.'[10]

In terms of the overall WMI Accord, one representative of the USWA, Richard Boyce, felt that it was very successful and there was 'a lot of growth on the part of individuals ... I honestly thought that I had much more in common with everyone else in the way in which the world turns after the process than before.'[11] Some labour representatives considered it worthwhile to sit down at the table with representatives of management's interests and talk with them outside of a formal bargaining setting. They thought that it helped to improve communications. Nevertheless, they, as are the other stakeholder groups, are waiting to see if industry representatives are prepared to buy into the whole spirit of the WMI rather than just that which promotes its own immediate competitive interests.

The Government Perspective on the Whitehorse Mining Initiative

> Canada's mines ministers recognized that a cooperative approach involving all stakeholders held the best promise for addressing the complex challenges facing the mining industry and for maintaining a prosperous industry committed to sustainable development and the creation of skilled jobs for Canadians. We were pleased that the mining industry took the lead in proposing this type of approach through the Whitehorse Mining Initiative. We were especially gratified that labour, Aboriginal peoples, and environmental groups came forward as active participants in the process. This collaborative approach will help to guide governments as they develop the economic, environmental, and social policies and programs that affect this important industry.[12]

Government perspectives on the WMI have already been discussed in the preceding chapters. For the most part, those who participated were fairly pleased at what was accomplished in terms of achieving contacts throughout the country and within their own governments. William McCann, director general of the Mineral Strategy Branch of the mining sector at NRCan, had responsibilities for coordinating the federal implementation of the WMI. McCann thought that the WMI was very successful:

> It is not easy to have multi-stakeholder consensus ... People should be quite proud. It is no small feat. In terms of follow-up by stakeholders, including governments, we should not expect things to be resolved quickly just because people have signed on to it. Now we are concerned about how to address the WMI goals and that's where we are now – a longer-term process. People are taking the WMI Accord seriously.

> The WMI Accord has been very important for us. It has helped us establish goodwill amongst other departments; participants and issues such as the environment fell under their jurisdiction and so gave the mineral sector more legitimacy with other federal departments – it showed that we could understand their problems and their constituents ... It also helped to focus the policy debate. It was very helpful from a public administration point of view.[13]

McCann also pointed out that the WMI is attracting attention in other parts of the world. He proposes that it can serve as a 'precedent' for the sustainable development of minerals and metals, and it has value in large resource-developing areas around the world. Although the Canadian model may not fit everywhere, the principle of bringing stakeholders together is very valuable.

Graeme McLaren, a manager of land-use policy in the mines ministry in British Columbia, is very experienced with multi-stakeholder approaches and was involved actively in BC's CORE initiative (see Chapter 6). He suggests that these processes are useful in nudging people from their original positions to more conciliatory ones. He does emphasize, however, that the processes are most successful at education and communication. One cannot expect major shifts. There will always be some who are not committed to the process of seeking agreement through give and take. From a government perspective, McLaren suggested that it was valuable to have the opportunity to meet people across the country who have their own understandings of what such terms as 'protected area' mean to them, or to get an understanding about how they have handled similar challenges in different ways.[14]

In his review of the Whitehorse Mining Initiative, assistant deputy minister of Ontario's Ministry of Northern Development and Mines, John Gammon, stated that the WMI was quite successful, noting that 'all participants were exposed to starkly different points of view and arrived at a much higher level of understanding.' Gammon also pointed out the value of the WMI in helping to develop the credibility of the mining sector with other ministries:

> It gave governments an agenda for action – recommendations gave ministers of mines credibility in other ministries. We could point out to other ministries that their stakeholders were also involved, including NGOs, environmental groups, Aboriginal peoples, etc. The WMI gave it legitimacy. It gave ministers of mines a mandate to advance these issues on a multi-stakeholder front. It gave them some leverage and so was very valuable.[15]

The Environmental Community's Perspective on the Whitehorse Mining Initiative

> Given that industrial development will continue, the environmental community believes that government and industry must adopt and strengthen programs, practices, and regulations to ensure that mining in Canada has as little environmental impact as possible. The Whitehorse Mining Initiative has the potential to be a critical step in removing the institutional barriers to putting the mining industry in Canada, and Canadian society as a whole, on a course to a more ecologically sustainable future.
>
> The environmental community chose to participate in the Whitehorse Mining Initiative to learn about the concerns of other stakeholders, to share its convictions about the environment and educate other stakeholders about the basis for these convictions, and to accommodate and support the needs of all stakeholders within the limits of these convictions. We seek from the mining industry and government support for three key measures of environmental sustainability: natural areas free of mineral development, better environmental practices during all phases of mining development, and more open, transparent, and fair decision-making processes.
>
> Government and industry are working to revitalize the mining industry, to address growing international competition. The environmental community maintains that there is also a need to address the nation's growing ecological deficit and to cooperate globally to protect our planet's stressed ecosystems. The drive to strengthen the mining industry cannot be at the expense of Canadian environmental objectives, health and safety standards, and the aspirations of Aboriginal peoples. The environmental community believes that, as a society, we must learn to address our social and economic needs within the planet's ecological limits.[16]

Members of the environmental community were pleased to have the opportunity to educate the mineral sector about their perspectives, as well as to learn about mining. Many of those interviewed, however, did not expect to see major changes in the mineral industry's attitudes. Some representatives voiced their unhappiness about the lack of follow-up to the principles 'as a package' underlined in the accord. One representative, Bob Van Dijken, from the Yukon Conservation Society, stated his feeling that the mining industry in the Yukon does not buy into the spirit of the WMI, nor does it recognize its relevance to the North. He adds:

> From our perspective, the key problem with implementation has been getting buy-in from industry and government. The Yukon Council on Economy and Environment [YCS] held a sectoral review on sustainable development in the mining industry in January 1995. Despite YCS asking

> that WMI be included as a topic for discussion during the session, this was turned down. We offered to do a presentation on the WMI, this was turned down ... The Yukon Territorial government also signed the Accord but has made no moves to implement aspects of the Accord other than those which conform to the Keep Mining In Canada agenda ... There seems to be a reluctance to deal with the other aspects of the Accord ... There must be balanced implementation of all Principles.[17]

Alan Young, a member of the BC Environmental Mining Council, has voiced similar concerns. He, like other members of the environmental community, saw the opportunity for mutual education. On the one hand, the accord was successful in that it gave the industry an appreciation for biodiversity and they were surprised to learn about the breadth and depth of the environmental issues. The long-term success of the accord, Young suggests, remains to be seen. Some members of the environmental community have the perception that industry has stepped away from the WMI commitment. Young referred to a questionnaire that was distributed by the BC Mining Association, which quizzed potential candidates for the next provincial election campaign. Among other things, the mining association asked the candidates which they thought was more important: economic or environmental protection. Young suggests that this type of question demonstrates that industry has yet to internalize what is meant by sustainable development and biodiversity. Decisions can no longer be determined on the basis of supporting either the environment or the economy. The two are inseparable. The industry responded that they simply wanted to know whether potential candidates were committed to mining or not and that this was not a sign that they did not support the WMI. Nevertheless, Young and others in the environmental community are concerned that the principles of the WMI have not permeated into the industry. He suggests that the WMI went farther than the industry expected but not as far as the environmental community hoped.[18] Ed Mankelow, of the BC Wildlife Association, was encouraged by the WMI process and its vision. He maintains that signatories to the WMI have to keep working through such hurdles as the one to which Young referred and must keep their commitments. All participants have an obligation to stay at the table and work things out.[19]

The Implementation of the WMI Accord

A little more than one year after signing the accord (on 23 November 1995), a meeting was organized by NRCan in Ottawa. Progress reports were presented on the actions taken by different stakeholders. It was generally acknowledged at this meeting that the hard work is just beginning. Invitations were sent to people who had participated in the WMI at any of the group levels – leadership council, working group, or issue group. Again, Dan Johnston facilitated the meeting of approximately fifty people – representing government, industry, environment, labour, and Aboriginal peoples. During the introductions, it was observed that: 'No one would have thought that all our problems would be solved on signing the accord ... [We have seen] the conclusion of phase one, and now the hard work is just beginning. We share a continuing need to educate others.' The stakeholders who attended the meeting gave a brief overview of their progress reports. It was clear that some progress had been made by some of the governments. But the verbal overview offered by the environmental representative was meant to 'cast a little bit of West Coast rain,' in his words, on the positive reports from governments that had been heard. He stated:

> Reading through the official responses one could take some comfort by the progress that has been made ... and I think some people in the MAC worked hard to spread the word. In British Columbia, progress has been made where many local activists embrace the WMI principles ... [There are] other regions where the WMI is still unheard of ... The WMI vision is vital to the environmental and economic health of the country. We do have the opportunity to avoid conflicts faced in forestry. People are trying to work toward an integrated vision and we all have a long way to go in each of our regions.

It was observed that the contemporary political climate in Canada could undermine any strides that governments had been making in the area of achieving the objectives of the WMI. This could include the potential loss of meaningful environmental monitoring because governments are downsizing, leaving the onus on industry. The industry, the environmental representative told the group, consists of 'shareholders, not stakeholders.' Furthermore, concern was expressed about a movement by provincial governments toward more support for business and deregulation. Since the WMI Accord, some of the stakeholders had a perception of an increasingly aggressive industry that did not seem as committed to the WMI Accord, particularly in the changing political climate.

The concern about the attitudes of new provincial governments in power (with particular reference to the new Conservative government in Ontario) was echoed by the labour representative who expressed a concern about the availability of 'government funding for training when they have a slash and burn mentality ... At the moment the atmosphere for labour management in Canada is at a bit of a low point, and that can't help but affect the way MITAC proceeds; we're working very hard to prevent that.' However, he concluded that 'everyone in this room has a huge investment in the WMI – some financial, some psychological – and everyone brings a commitment to devote significant energy in the future. It is important to recognize the problems and address them with a sense of realism.' He added a reminder: 'Industry has a significant stake in not losing momentum. It is tempting with a change in government to think that it will sing from your song book and to go for it with gusto, but that is a short-term strategy, with some long-term pain associated with it. There are some warning signs on the horizon for the process we all invested in.'

An Ontario representative followed this discussion with an effort to assuage some of the concerns about the change in government. He acknowledged that the government wants to decrease both the budget and the size of the bureaucracy. He stated: 'We understand that there's a lot of pressure to retain high environmental standards ... we see very large challenges to make these changes with fewer dollars, fewer people, and fewer regulations. However, the [provincial] government is aware of the WMI. It realizes the industry supports it and that this is a multi-stakeholder process ...' An industry representative from CIM pointed out that 'there are people out there working in remote areas who love the land, protect it. They are the ones who will carry out, monitor, and self-regulate and test. There's a good solid base of understanding and trust. There is a lot more work for trust needed, particularly under self-regulation ... CIM promotes the process and our members are more than willing to carry it out.' From all sides it appeared that there was a degree of goodwill to continue on with the WMI. It was also clear, however, that there were some important challenges to be addressed and that continued communications and discussions were vital to the process.

Participants at the seminar were asked to address a series of questions and points in discussion groups (see Table 7.1). Discussions in the workshops were wide-ranging. Deliberations in one workshop focused around five main areas. The first was a concern about a changing, more conservative political climate that might serve to undermine the goals of the WMI, such as maintaining high environment and health and safety regulatory standards. The second theme examined the difficulty in ensuring that the principles of the WMI would be widely communicated and adopted throughout the various jurisdictions and diverse stakeholder groups. The third theme involved a discussion about encouraging the mining sector to comply and work voluntarily toward the goals expressed by the WMI Accord. Fourth, the question of accountability and compliance was addressed. If the ideas of the WMI were successfully disseminated and adopted, how could progress be assessed? Finally, the issue of partnerships and multi-stakeholder approaches to mining policy was considered.

In the workshop, one labour representative introduced the question about the success of implementing the goals of the WMI in the current political climate. He made several perceptive observations:

Table 7.1

Working group agenda

(1) What new policy initiatives or policy changes are necessary to assist in the further implementation of the principles and goals in the WMI Accord?

- National harmonization of standards/methods regarding:
- environmental assessment and permitting
- integrated land-use planning
- labour force development
- mine reclamation policies
- completion of protected areas (national, provincial, territorial)
- completion of comprehensive/certain mine reclamation policies.

(2) What actions can governments and/or stakeholders take in partnership to further implement the principles and goals of the WMI Accord?
- national industry/environmental workshop on protected areas and mineral development
- regional multi-stakeholder workshops on mineral exploration and development in sensitive areas
- sector councils (e.g., MITAC) to develop training and adjustment strategies
- 'Model Mine' projects.

(3) What specific actions can governments and/or stakeholders voluntarily take to assist in the further implementation of the principles and goals in the WMI Accord?

The summary of status of the WMI implementation provides a brief overview of some of the future actions currently being contemplated by various governments and/or stakeholders. It is suggested that this part of the discussion be used for stakeholders to:
- generally/informally describe some of the voluntary action that they will be taking in the future
- suggest voluntary action that could be taken by other stakeholders that would significantly assist in further implementation of the WMI Accord.

(4) What specific actions can be taken by governments and stakeholders to increase/promote better understanding between stakeholders and to increase 'constructive' engagement between stakeholders?
- more direct consultation and communication with communities
- workshop identifying what creates 'distrust' between stakeholders and how this can be addressed
- proactive conflict resolution or facilitation at policy level and at project level.

> The question in my mind is whether the WMI, in effect, is an accident of a particular moment in history when at least two or three major provincial mining jurisdictions had governments with a particular perspective, soon to be followed by another ... When the context changes does the gas drain out of the tank or does the WMI philosophy or approach have a life of its own? I haven't answered that question for myself ...
>
> It is a pretty stormy climate out there for the kinds of things that the WMI stands for. I don't think you can answer the [four] questions about harmonization, protected areas ... without addressing questions about changes in context. Lots of things on this list [see Table 7.1] assume that there will be financial resources from government to smooth processes along. It presumes there's an atmosphere for a more efficient, streamlined, pragmatic process but one that is publicly accessible, reproducible, and transparent.
>
> On top of that you have pre-existing and bigger question marks about the role of the feds or the role of national standards in this rather strange federation which operates in two senses – constitutional and fiscal. As a union we have to struggle not to import feelings of outrage that we develop in other aspects of our economic and political life in these forums and recognize that there's some give and take ... we have to struggle to keep an open mind. Industry has the toughest decision to make because there's no question that the political environment in Ontario, and potentially in other jurisdictions, is very favourable to the kinds of things industry has on its wish list. Industry has to decide whether to go for the gusto or to go with the longer vision. My experience, relative to other sectors, is that industry tends to take a very long view of the world ... Industry has to decide whether ... it wants to step back and see a longer view and *reinvest in the process*. [our emphasis]

George Miller acknowledged concerns expressed by other stakeholders about the changing political climate in provinces such as Ontario, with the emphasis on reducing the deficit in ways that may undermine the work of labour and the environmental community. He also noted the difficulty that each stakeholder around the table had in terms of his/her limited ability to speak for his/her stakeholder group: 'Industry is not a monolith. Whether [we] ... can bring solidarity I don't know ... There are some political forces that will help move the WMI in some ways, and make it harder to maintain or balance in other ways.' Another participant offered a perspective about the financial restraint plans of governments throughout the country and its potential impact on the WMI: 'We have a golden opportunity today as influence groups to make a statement, a reaffirmation, to step up efforts at the provincial and territorial levels ... Because of the fiscal reality, we may be driven to partnership. We don't have the finances to lubricate the process, but it does generate some creativity.' Certainly, the financial constraints evident throughout public sectors in every jurisdiction will make it more difficult to implement the principles of the WMI, such as improving training opportunities.

There are specific concerns across all sectors, not just the mining sector, that the direction of change in the regulatory environment merits consideration and even, according to some observers, a degree of concern and caution. In the October 1995 edition of the *Financial Post* magazine, the potential pitfalls of Bill C-62, the Regulatory Efficiency Act, were briefly explored. In response to an overly complex, sometimes contradictory regulatory environment, the proposed law would allow companies to negotiate their own compliance plans, rather than adhere strictly to the existing regulatory framework. Critics worry that 'The Liberal government, in its haste to unburden corporate Canada, has just divided federal regulations pertaining to business into two categories: One for the rich and powerful (them) and another for the struggling and powerless (you) ... Bill C-62 will create a regulatory no-man's-land and tilt the competitive playing field against the already beleaguered small business sector.'[20] Environmental and labour advocates are also concerned about the impact on health, safety, and the environment if governments move toward voluntary compliance measures rather than regulations that require more rigorous monitoring by government. To such concerns, Miller responded that 'regulatory simplicity can expect to find sympathy [within industry] but if we see increasingly low [environmental] standards we'll all be the losers in the long run.'

Along with expressed concerns about an inhospitable political environment in achieving the goals of the WMI, participants noted the uneven reception that the national initiative was receiving in provincial or regional jurisdictions. Bob Van Dikjen, from the Yukon Conservation Society, noted: 'In the Yukon and with DIAND there doesn't seem to be a leadership role in fostering a regional process. We have run into a brick wall at all levels in trying to go from a national to a regional process. We need the government level to foster discussion, to have working groups such as this. The government role has not been followed up at all ... We need to foster subsets of the national WMI.'

Problems in other regions were also noted. Serious concerns from Aboriginal peoples in Labrador were expressed. The group was reminded that Newfoundland was not a signatory to the WMI and that the government was 'opening the door to all kinds of development ... For us the prime way to implement the WMI shouldn't only be government's responsibility but also the mining industry's. We're dealing with a lot of junior companies involved in exploration ... none has ever come to an Aboriginal group ... only Noranda is very good.' Throughout the day, it was agreed, as George Miller stated, 'that the responsibility for getting the word out to industry should be made very explicit.' If the goals of the WMI Accord are well communicated, they could influence the operations and procedures associated with the many facets of the mineral industry. It may prove to be particularly useful in encouraging voluntary initiatives in the mining sector. George Miller suggested that a management framework might be established to support efforts to carry out policy; 'it wouldn't be detailed but [such a framework] could be our next step in the evolution ...' The framework might include some direction about the kind of incentives, training, and other efforts that are needed in trying to fulfil the WMI vision. Brian Parrott, of the BC Ministry of Energy, Mines and Petroleum Resources, pointed out that some voluntary actions on mine clean-up were taking place in northern British Columbia; for example, 100 old barrels of fuel and various pieces of old equipment were taken away in a cooperative arrangement between the industry and the ministry. A federal government representative stated that 'voluntary approaches, particularly in the environmental area, is an idea whose time has arrived.' Incentives and disincentives become even more important in the context of regulatory changes toward voluntary action. For example, in an effort to encourage 'eco-efficiency,' one could introduce a deposit and return system for oil drums rather than a regulatory framework that penalizes a company for leaving an oil drum in the bush; a $50 deposit encourages the company to return it.

In considering ways to encourage voluntary industry compliance, it was suggested that professional engineering societies could be used as de facto agents of the public to ensure that reclamation is being undertaken; consideration was given to whether engineers and geoscientists could be 'deputized.' In a movement toward voluntary, internal responsibility – motivated not because of a positive track record of those who are being regulated, but due to the need for governments to reduce their expenditures – issues of accountability and credibility must be considered carefully. It may be that to enhance the credibility of internal responsibility systems, industry will have to invite people from outside to become involved in the accountability process. Auditing processes are another or a parallel route to ensure that standards are maintained. George Miller suggested the adoption of the International Standards Organization (the ISO was designed to ensure international quality assessment standards were in place and there is movement toward establishing international environmental assessment standards) certification processes. In particular, the most recent ISO 14 000 certification series would enhance the industry members' credibility and accountability practices; the ISO 9000 and 14 000 series involve an extensive creditation process and periodic audits. He suggested that 'a company that qualifies for ISO 14 000 will have made a huge investment and will not want to jeopardize that.' Environment Canada is working to support companies that have gained ISO certification. Such companies, for example, will be audited periodically, but may be relieved of the twenty-times-per-day stream sampling for ensuring effluence control. These standards may increase the level of confidence in the industry and its credibility. Certification would also likely increase the banking sector's willingness to lend money. Indeed, someone suggested that the 'WMI could become an "ISO" itself,' if only at the national level.

The real test of the WMI will be how well its principles will be respected when there is a rush to open up a new area to mining, as in the case with the Broken Hills Property (BHP) diamond mine in the Northwest Territories or with the new mine in Voisey's Bay, Labrador. The early response has been mixed. In the case of Labrador, serious concerns have been expressed about the tension between prospectors accustomed to free access and the lives and interests of Aboriginal communities. With regard to the BHP mine, reservations have been expressed about the limitations of the environmental assessment process. Critics have noted the lack of time and resources available to the review panel as well as unsatisfactory treatment of issues involving monitoring, site reclamation, lack of use of traditional knowledge, and community consultation.[21] On the other hand, the BHP mine will bring jobs necessary to the mixed northern economy, which combines community-based subsistence activities with wage- and income-based activity.

On 1 November 1996, the $900 million BHP project received full government approval. Several agreements are pending, but one was signed that focuses on employment and training opportunities for northern Aboriginal communities. Reservations are still expressed about by some members of the Canadian Arctic Resources Committee, who have pointed out that the question of land ownership and treaty entitlements for the Dene have yet to be resolved. Furthermore, they expressed concerns about the adverse environmental impacts on some northern lakes and the migration paths of the Bathurst cariboo herd. An independent committee is expected to monitor the project in order to provide environmental safeguards.[22]

Communicating the WMI throughout the different constituencies remains a challenge. This is particularly the case with mining companies, particularly junior companies. Miller stated: 'I accept that industry has to get the message out to 3,000-odd mining organizations out there. I accept the challenge.' One government representative stated: 'We have to try to embody the principles and the mindset of the WMI into the processes, procedures, and relationships with all the various stakeholder groups. We need to push this and regionalize it more than has been.' In each stakeholder group, leaders such as George Miller are necessary for such initiatives to be successful. This applies both to the facilitation of the initial stages of the WMI, as well as to making the commitment to carry the message to others in their communities. Throughout the meeting, there were numerous ideas for disseminating and developing widespread support for the goals of the WMI Accord. It was suggested that an internal process could be established for each stakeholder group, such as the professional associations, to encourage or require its members to 'sign on.' The difficulty, of course, is that because the general members did not participate in the WMI, they may remain unpersuaded as to its merits. Nevertheless, many felt that the suggestion was worthwhile. The CIM, for example, has 12,000 members and sixty branches. If the national association adopted the WMI principles, and was successful in disseminating the ideas to the branches, progress could be made adopting and taking action on the ideas expressed in the accord. It was often suggested that the accord and the work of the issue groups should be made readily available and be broadly distributed. It was noted, for example, that a copy of the WMI Accord should be issued when someone applies for, and receives, a prospector's licence.

It was clear that working group members around the table believed in the vision of the WMI and were searching for ways to share the vision. It was agreed that care must be taken to ensure that some members of stakeholder groups did not polarize or over-simplify policy choices; leadership within the different stakeholder groups should discourage such action that leads to confrontation rather than cooperation. One public official in the group maintained that partnership is critical; this is what is needed if the goal is to move away from advocacy processes and move toward partnerships in processes such as environmental assessment.

The issue of accountability was addressed in the context of ensuring that the principles of the WMI are acted upon by all the stakeholders. It was agreed by the group members that what is needed is a score card to assess policy results across the country. Hugh MacKenzie, a labour representative, observed that the lobbying strength of non-government stakeholders varies from jurisdiction to jurisdiction; 'we don't know if something is going off the rails in Voisey's Bay, for example.' Alan Young recommended that the mines ministers conference be used as a vehicle for assessing progress – 'for comparing report cards' – on the WMI Accord. It was suggested that a template be developed – a score card – that would allow stakeholders to report, in a uniform way, what progress was being made toward realizing the WMI principles and objectives. One government representative observed that the ministers' conference had evolved in the past few years from a process that was very closed – even to industry – to one in recent years that sought to engage a more diverse range of interests in the meetings. At one session during the most recent mines ministers' conference, an industry representative sat beside the Saskatchewan minister; there may be an opportunity to bring other stakeholders into some part of the meeting. One concern was that the WMI Accord cut across departmental boundaries, which meant that other ministers, such as environmental ministers, should be involved in the reporting progress; it was suggested that in some way, the councils of ministers should be brought together. In addition, signing the accord was recognized as a way of taking ownership over the principles and recognizing that the WMI Accord sets out the expectations that have to be fulfilled.

Communication throughout the membership of non-governmental stakeholders, such as the environmental movement, remains an issue. Alan Young stated: 'There are those people who think that it [the WMI] was a sell-out. They are sceptical about a non-binding Accord. We've argued with them that these are a set of principles that we can work with. It [the WMI] is taking root amongst those who are interested in dialogue. The full spectrum, from negotiation to protest, is to be expected. I believe it is necessary and useful to have diversity if you want to gain progress on environmental issues.'

The question of partnership identified in Table 7.1 was also addressed. Given that there is some lack of understanding about the scientific basis for establishing areas for protection, it was recommended that a small workshop be offered that would fill some of the information and knowledge gaps. Miller stated: 'Within industry there's some scepticism about whether the extent and location of proposed protected areas needs to be, how big such areas should be. Industry might be persuaded if we asked scientists to explain the foundation for such judgments.' It was recognized that there are some outstanding issues that can be addressed, and possibly resolved to the stakeholders' mutual agreement, by adopting the environment for learning and the appropriate setting for dialogue, which was initiated by the WMI. One challenge to be faced is how to finance the participation of stakeholders in the future workshops and partnerships. Government has been an important sponsor of multi-stakeholder efforts; however, with fewer dollars available from all government jurisdictions, the challenge is how to get people to the workshops. Hugh MacKenzie from labour stated: 'We find it very difficult to participate in all the consultative processes we are asked to participate in, and the situation is worse yet for environmental groups.' Industry representatives pointed out that companies have in the past provided funding to ensure that, for example, Aboriginal representation is possible. Although consultation processes are costly, they can be valuable. Alan Young noted that while participants in the workshop might arrive at some level of mutual understanding, it is difficult to translate that to one's constituency. He recommended that a video of the workshop be produced 'so that others can see the dynamics of what went on.' It is possible that the challenge of communication can be addressed by using technology to help disseminate information and communication. In a visually predisposed society, it is true that people are more willing to work through a video than they are to plough through written pages of a manual.

Recommendations for a 'Model Mine' project stimulated considerable discussion as well as raised some concerns due to its vague meaning. One could interpret that the concept borrows from the model forest developed in the context of the National Forest Strategy, but there was some scepticism about its applicability to the mining industry. Another way to interpret 'Model Mine' is as a showcase model. Such an interpretation would reflect the 'leading by example' approach. One participant suggested that such model mines exist; for example, certain environmental programs at the Highland Valley Copper Mine, such as oil recycling, stepped reclamation, and whistle-blower protection, should be looked at as important steps forward. Miller agreed that showcasing could be done: 'It wouldn't hurt to start small ... we could get some people together to discuss how this could be done.' In northern Saskatchewan, for example, one could look at the work of the Northern Labour Market Committee, which includes representatives from a number of communities, working collaboratively in addressing common labour issues. One industry representative pointed out that there is a small mine he knows of that has been planned with closure in mind. A working group was established there to help workers design their own work routines. Another example was offered of a mine that has undertaken 'some serious scientific work on acid drainage; all the lessons from the Acid Rain Research Program were incorporated there, ... and the mine got its permit in record time.' The term 'model mine' is a good example of the need to pay close attention to language. 'Voluntary showcasing' of good practice by a particular mine – without creating an unduly competitive context (undue given the different resources and other unique variables each mine faces) – was agreeable to the group's members. One mining official said that his company has found it beneficial to establish good practices so that when one wants to establish a new mine one 'doesn't have to say, this is the way we would do it. Instead, one would say "Look over this hill, this is *my* standard, not industry's."'

Conclusions of Post-WMI Meetings

Several issues were raised at the November 1995 meeting. One primary concern is whether there exists a WMI champion or ambassador; there was a recognition that the WMI vision could not survive in a vacuum and that some kind of an implementation team was needed. One proposal was that there 'has to be a more concrete and active role for the federal minister.' The WMI should not be seen as an end in itself but rather as a process, as a tool. Within the ranks of the stakeholders, too many people do not know what the WMI is about; therefore, there is a need to 'communicate and sell through our ranks using paper, video, radio, speakers, and translation into Aboriginal languages.' The annual mines ministers' conference could be the appropriate place for the WMI to be discussed as a *formal item* on their agenda. One observer said that there was 'nothing formal [about the WMI] on the last mines ministers' agenda.' Other points that were brought to attention included:

- there should be a workshop on how to work together for greater certainty on protected areas
- public consultation on mining development has to be early and public
- intervenor funding is necessary and the MAC should convene a group to discuss criteria for such funding, including how much and to whom funding should be available
- there should be a workshop on how to deal with voluntary self-regulation versus government regulations
- what is needed is not only educating people about the WMI but getting companies to actually adopt it
- a process to continue the WMI with a shared agenda in what is becoming phase two is needed
- people can participate in the WMI in different ways, using technology such as teleconferencing
- there is a limited awareness of the WMI – it needs to be communicated throughout Canada, including our colleges and universities
- there is a need to develop a coordinated WMI position in support of environmental assessment processes
- there is a need for a land-use planning process and a need to bring other stakeholders into the process, such as oil and gas people
- there is a need to bring land-use and management concepts together
- there is a need to support and expand MITAC and a need for an interprovincial 'ticket' for graduates (a national skill basis is required to ensure mobility of employment).

The commitment to the WMI vision was quite evident, although it was also clear that there are some challenges. Some of the most pronounced were with the concerns of Aboriginal peoples in Labrador to have the WMI principles and objectives recognized and acted upon by junior companies. There were also concerns in the northern territories, where jurisdictional questions and organizational cultures unique to the northern mining communities were blocking progress on the WMI Accord.

The November meeting was clearly needed – after a year away from the process, there was an opportunity to reflect on the high degree of consensus that was achieved and the ongoing commitment to fulfil the promise in the accord. It was clear, and not surprising, that there exists a degree of fragmentation within the different stakeholder groups and that work needs to be done to legitimize the accord. At the end of the meetings, George Miller, who spearheaded the whole initiative, concluded:

> The onus is on the people who are involved in making decisions that affect many other people so some optimum can be realized in the social, political, and economic sphere. I pledge my commitment to move the WMI principles down through my organization. Some real world changes are to be expected. I hope every other group will do the same. I hope that this first post-signing reunion will not be the last, and that we will have other meetings to review opportunities and constraints.

As is often the case with such initiatives, the problem will be maintaining momentum. The short-term success of the WMI was acknowledged. It served a badly needed educational function, diverse groups were able to come to an initial consensus on many important interests, and the accord may be used as a blueprint for future directions in the formulation of mineral policy. The long-term success, however, is still unknown. It is to this question that we now turn.

8

Terra Incognita: The Future of Resource Policy-Making in Canada

> It is always possible, and sometimes exciting, to escalate debate to the point of fundamental principles, basic rights, and non-negotiable demands. But the lesson surely is that the upshot is seldom increased understanding, let alone agreement. Much less dramatic, but much more promising, is a discourse that searches for more mundane accommodations.[1]

In spite of the relative success achieved with the signing of the WMI Accord, there are notable challenges ahead. Similarly, there are exciting possibilities for regulatory reform that could offer a viable approach to sustainable mining. Canada has been a world leader in many aspects of mining. It could be so in this arena as well. The opportunities and challenges include extending and supporting the role of advisory councils and similar initiatives, constructing some tangible signs of accomplishment and reward, international agenda-setting, continuing a yearly report card on the progress toward WMI goals and initiatives, and ensuring that institutional mechanisms are in place to promote and sustain such initiatives.

Opportunities and Challenges

It is quite clear that the WMI would not have happened without the breadth of vision and commitment shown by George Miller, president of the Mining Association of Canada (MAC), and other members of his association, as well as other key players from all the participating sectors. Those people not only had to devote a tremendous amount of personal energy to the process, they also had to demonstrate to others less committed to stay with the process. It was the force of their personalities, dedication, and ability to persuade others that made the signing of the accord possible.

The degree of commitment will be much more difficult to sustain over the long term. Leaders of such processes need to have reinforcement over the long term. Long-term commitment needs to be present in significant numbers in all the affected communities of interest.

Participants' perspectives on the accord were wide-ranging. At one end, there were those who were cynical about the motivations of the other stakeholders, detached from the process, and who rejected the accomplishments of the WMI, either at the time of signing or in the following year. At the other end of the spectrum were those who seemed to internalize the value of enlightened self-interest. Members of this group served to stimulate and inspire others to compromise and honestly attempt to come up with a consensus-based decision. Those people who are able to come to the table relatively unencumbered by the expectations of their diverse communities will be most able to participate fully in the potential offered by roundtable consultations.

Long-term success of the WMI will depend on the degree to which those who showed leadership in the initiative will be stimulated into ensuring that the process will continue into the next stage. The question of how this interest should be fostered and by whom is a difficult one, because those institutions with money and resources will have to be willing to support these individuals in their initiative.

Can the spirit of the WMI and the commitment to its principles be sustained? Some of those who had very high expectations failed to see that this effort represented the very beginning, not the end, of a long journey in so-called 'shared decision-making.' To ensure long-term commitment, participants need a sense that their interests are being protected and promoted. This is where advisory councils can play a very useful role. If multistakeholder advisory councils were established in every province, and were given sufficient resources to meet and communicate regularly, it would be possible to develop a regulatory regime that would better serve everyone's interest. If members of the council were respected for their knowledge and ability to be broadly representative, they could play an important role in ensuring that legislation was developed in a way that would accommodate a variety of concerns.

To begin with industry, mineral representatives would like to be allowed to be more self-governing and they would like to have a more predictable regulatory regime. An increased measure of self-governance could take place if an advisory council was in place that was active and alert enough to monitor the legislative changes and ensure that environmental concerns, for example, would not be compromised by such changes. Only an advisory council that had a visible public profile and broad enough representation would be able to serve in that capacity effectively.

An advisory council could also serve the interests of government by streamlining the consultative processes. Instead of continually dealing with a variety of interests on a project by project basis, consultation could be institutionalized in a way that an advisory council could deal with broad policy concerns ahead of time. Consultation, then, would need only affect the immediate community in which a specific project was to take place. An advisory council would also ensure that Aboriginal perspectives were represented, although this would not take the place of government-to-government negotiations or negotiations with individual communities.

Environmental and union concerns would very much benefit from the creation of a strong advisory council, because they would be at the table at the policy-formation stage rather than left in a reactive position.

The council could also play a role in public communications, education, and networking. A considerable amount of the work and public communications could be done through electronic mail, the World Wide Web, newsletters, etc. Participants would volunteer their time. It could also serve as a vehicle to monitor the progress each government has made in terms of the WMI objectives. It could then advise the ministry on issues that may need attention. Furthermore, any activities that various mining companies demonstrate a particular form of commitment to could be highlighted and publicized by the council. This recommendation would particularly be in the interests of mining companies since investors are risk averse when resource companies become the subject of unfavourable public attention.

To support such an initiative, the various ministries would need to underwrite some of the expenses of the enterprise. It would be in the interests of industry to support some of the public education functions. For those who point out that this would be an expensive undertaking, particularly in an era of downsizing, one can easily proffer a number of counter-arguments. First of all, participants' time is provided on a volunteer basis. Second, the council can serve to legitimize the decisions of the government and help provide a regulatory and monitoring function. Third, the council can play a proactive role in the policy-making process and prevent the very real and expensive delays at the project-approval stage because of failures to accommodate a broad diversity of concerns. Fourth, mining revenues are important to the economy. If an advisory council can help ensure that mining is carried out in a responsible and sustainable way, the regulatory environment should become more stable and attractive to investors.

The difficulty here is demonstrating to governments and industry that these initiatives are not a waste of time. More leadership will be needed to accomplish this task. Governments have yet to figure out how to effectively make use of public consultation and are unwilling to provide too much direction for fear of being accused of trying to direct the results. Industry needs to be shown that it is in its interest to support the councils. Definitive goals need to be established.

The Whitehorse Mining Initiative can play a very important role internationally. Application of principles based on the WMI can lead to the establishment of national and international standards, not only in terms of environmental assessment but in areas of health and safety and community responsibility. Companies could receive certification that acknowledges that they are meeting these standards. International recognition of this kind would encourage investment in companies practising sustainable mining using the best available technology and the highest standards. Other countries that are active in mining, such as Indonesia, South Africa, and Bolivia, have been watching the WMI and are requesting seminars about how similar initiatives may be applied in their own domestic setting.

The WMI represents a considerable opportunity to set international agendas and show leadership in this field. The selling of Canadian expertise abroad and the establishment of international standards of mining practices would be an impressive accomplishment and should act as a considerable incentive. Supporting the initiative would be in the interests of all Canadians, industry, environmental groups, northern communities, unions, and other actors and agencies.

Effective communication of the WMI principles is a very important responsibility that the signatories of the national and provincial statements of commitment must continue to play. There are large numbers of people in the various policy communities who did not attend the WMI. These individuals did not benefit from the learning process, did not understand why certain compromises were made, and remained distrustful of the process. The leaders of these groups and organizations who signed the accord will have to do a good job of educating their own constituencies if they expect support for the WMI vision. There is also the temptation for the participants themselves to slip back into their former positions, when memory of the WMI fades and more recent concerns replace it.

The actual short- and long-term influence of the Whitehorse Mining Initiative is contingent to a great extent on positive impressions and awareness about the goals of the accord. As is so often the case with the mining industry, however, the WMI received very little public attention. Communicating its goals and disseminating the message, even among its own industry stakeholders, has proven to be a considerable challenge. One senior government planner pointed out a letter to the editor of the *Northern Miner* to illustrate the challenge. A president of a northern prospectors' association had written a letter strongly expressing his concerns about the 'fluffy language and objective' of the WMI: 'We are starting to sound just like *them,* a sure sign our leaders are getting too cosy with the opposition.'[2] This kind of adversarial approach will make it difficult for the industry to promote the goals of the WMI and sustainable mining. The activities of those unwilling to compromise and recognize the new political environment will serve to undermine the industry's ability to influence the agendas of political decision-makers.

The media, for their part, have not played a very helpful role. Progress made in mining often goes unheralded. Art Ball, a government representative from Manitoba, points out that they have undertaken 'some magnificent rehabilitation work ... and no one knows about a lot of it ... The general public doesn't realize that the total lands used for hard-rock mining in Manitoba in the past 100 years is far less than the area of Winnipeg!'[3] Responsibility for the lack of public knowledge may lie, in part, with the media. At the first leadership council meeting of the WMI at the PDAC conference in March 1993, Manitoba representatives went to meet with the editorial boards of the *Globe and Mail* and the *Financial Post.* One provincial representative remembers: 'They scribbled no words. It wasn't news ... there was no element of confrontation!' Unless there is a confrontation between different groups – a mining disaster or some other dramatic event – developments in the mining industry do not receive coverage by the media. Space will be found in a newspaper if one can construct an enticing provocative headline. Even in the case where mining is its focus, the *Canadian Mining Journal* published the headline, 'The Whitehorse Whitewash.' From that article, the following quotation expresses the newsworthiness of the story – not the possibility of consensus, but of failure: 'Moreover, its [the WMI's] 2-year $1-million study of mining practices and policies within the country will produce *only* broad changes in strategy, rather than specific directives for those in the industry to follow.'[4] The article is not wholly negative but its headline certainly is – reflecting the media's predisposition for sensationalism. A supplement to the September 1995 *Report on Business* waved a headline, 'A Future Denied,' for its article about the Canadian mining industry's future.[5] The remarkable achievement in reaching a degree of consensus realized by WMI stakeholders with such varied interests, values, and knowledge bases is not exciting enough for a headline in the media.

Given this lack of attention and a degree of mis-representation in the media about the WMI, it may be that there is a need for WMI stakeholders to step-up their efforts to communicate the nature of the accord and the significance of the process itself. The eventual pay-off from the WMI process will be long-term and realized in ways not always predictable in advance. The WMI goals, principles, and processes need to be driven home at the regional and community level.

After the Whitehorse Mining Initiative: Implementation

Perhaps the most interesting issue that has yet to be satisfactorily addressed concerns what will follow these roundtable public consultations on sustainable development. Some individuals in industry and government may hope that there will be a return to a simpler day, when the public demands for consultation will not be so vociferous, and the numbers of interests that must be accommodated are more clearly specified. This appears unlikely. Regardless of the ideological orientation of the government of the day, public consultation is increasingly becoming a convention that politicians would be ill-advised to ignore. Those trained in dispute resolution should find themselves in a growth industry. Government officials will receive training in managing these processes along with their traditional duties. Participants may learn to become good listeners, communicators, and possibly consensus-builders. In a world where integrated resource management is a necessity, there appears to be no alternative. But the question remains – does it work?

The scope of changes in land-use planning currently in progress throughout Canada, particularly in British Columbia, are dramatic by global standards. The significance and ramifications of these changes for public policy, administrative behaviour, and representative democracy warrant careful scrutiny. This approach to decision-making is based on a new model of public representation. The public has traditionally been represented by elected representatives and parliamentary institutions. The interests of various constituencies have been integrated and articulated through the political process by elected representatives, parties, elected assemblies, and, ultimately, the executive.

In the past two decades, the growth of the role of the media and public interest groups, the entrenchment of the Charter of Rights and Freedoms, and so on, has led traditional forms of representation to be perceived as incapable of adequately integrating the diverse needs and demands of the population. It is common to see bureaucracy portrayed as 'leviathan,' the public service as unresponsive, and governments as incapable of responding to complex demands generated by the public and the complex machinery of public administration.

In British Columbia, where land-use conflicts seem unending and unresolvable, provincial reforms have been extensive. They are notable for the unprecedented scope of public participation, the comprehensive pieces of land-use legislation that have passed, and the implications of such changes that have yet to be adequately examined.

How, then, do we know that something is being achieved? We must first consider whether genuine reform is taking place. Are processes such as the British Columbia Advisory Council on Mining (ACM) really new? Are they changing the nature of decision-making and will they actually lead to some sort of administrative reform? The impact may not seem as far-reaching as one may, at first, assume. Advisory bodies, after all, are not the ultimate decision-makers. At the end of the day, that responsibility rests with cabinet. Moreover, advisory boards, councils, public consultation exercises, task forces, and royal commissions are by no means a recent phenomenon in Canadian politics, whether the issues pertain to the introduction of universal services to health care and social programs, Canadian communications, transportation, culture, or mega-project development. The difference of the new initiatives lies in their scope; public consultation today is undoubtedly much more comprehensive and sweeping.

It is also clear that the WMI Accord has not been fully endorsed across Canada:

> The provinces of Newfoundland and Quebec and the Assembly of First Nations were not able to sign the Leadership Council Accord at the time of the September 1994 Mines Ministers Conference. The province of Newfoundland was unable to sign because the Accord had not been endorsed by the provincial Cabinet [Newfoundland later signed the final poster]. In Quebec the government elected on September 12, 1994 wished to examine the Accord before deciding on a position about it. The province of Alberta would not sign the Accord. The Assembly of First Nations advised that widespread consultations had not been conducted with First Nations. Additionally, land claims and treaty groups had not been consulted to assess the impacts of the Whitehorse Mining Initiative. Most Leadership Council Members expressed complete support for the process as a whole.[6]

Such is the nature of our federal system of governance. Provincial jurisdictions that do not initially embrace the WMI will be watching other provinces to see what is working and to monitor the effectiveness of their efforts to implement the vision of the initiative. Federalism need not have a conservative or dampening effect on the initiative. Sometimes, different provincial approaches to implementation can be innovative (as was noted in Chapter 6) and can serve to encourage others to follow suit. This would not be a first for Canada; one only needs to consider that the formation of the national system of health-care insurance began in Saskatchewan.

What about the process of decision-making itself? How much influence do multi-stakeholder advisory groups have on the policy outcome and implementation? Much will depend on the success of the group in demonstrating cohesiveness and showing initiative. If this can be accomplished, it will most likely have an influence on the final outcome. If a consensus cannot be reached, governments are more likely to take over the process and impose a solution. Obviously, the more diverse and representative the group and the more comprehensive the issues, the more difficult it is to achieve a consensus.

We hear a great deal about administrative reform, but it is unclear how much is actually taking place. Government officials do find themselves interacting more intensively with a broader variety of interests than before. Advisory groups also serve to consolidate and streamline decision-making. When no consultative processes take place, interest groups react to individual project or development initiatives in a variety of ways. Governments have found themselves in the position of responding to those reactions on an individual basis, which leads to an unstable policy environment. Consultative processes, therefore, may integrate and coalesce those views and streamline the planning process. The rules would then be in place before a great deal of time and effort was invested in a particular project and before the stage is set for a land-use conflict. If a decision is subject to the scrutiny of a plurality of perspectives before it is implemented, governments may be viewed as more accountable than has been the case in the past.

Consultative processes will have an impact on the way that public officials carry out their duties. As members of the interested public are included in the advisory process, public servants will be increasingly judged or assessed on the basis of how responsive they are to client groups. The difficulty for the administrator will lie in determining the client group.

More open decision-making processes in the mining sector may or may not lead to significant improvements in the administration of public policy. Ironically, as efforts are stepped up to improve government accountability and responsiveness, governments are becoming more complex. Interdepartmental lines are blurred, and the lines of authority have also become increasingly muddled and confusing. This presents the public with the perception that governments are less accountable and responsive. As Evert Lindquist noted:

> Even if a thousand management flowers bloom in the public services across Canada – such as single-window service delivery initiatives, special purpose agencies, and diverse partnership arrangements – an interesting paradox will emerge: as a multiplicity of government entities and partnerships are created inside and outside the public sector, government inevitably looks bigger and more complex to citizens, even if government is getting smaller and the organizational innovations are effective and constitute authentic decentralizations.[7]

Various models have been proposed in recent years to widen the opportunities for participation by the public. We are now seeing a number of public roundtable exercises that have significant impact for the role of the public official and, more fundamentally, on our representative institutions.

In his insightful analysis of current public policy trends, George Hoberg suggests that Canada is seeing two new processes emerging: multipartite bargaining and legalism. While the two developments may be complementary, they are also based on very different ideas and 'world views.' The adversarial, formal characteristics of legalism are one approach where the courts oversee interest representation. In contrast, multipartite bargaining rests on ministerial discretion, an informal bargaining process that encourages consensus-building. Hoberg suggests that legalism is well suited to the American style of government with its separation of power, while multipartitism carries on the Canadian tradition of ministerial discretion. Nevertheless, with the introduction of the Charter of Rights and Freedoms, the hostility of the Canadian political culture to legalism may be declining.[8] He concludes:

> The Canadian policy style is clearly at a crossroads. Whichever road Canada chooses, it is important to realize that neither of the emergent policy styles addresses one of the fundamental problems underlying governance in this area; the weakness of the Canadian state ... While the state has been busy organizing consultative forums and resisting the encroachments of the judiciary, *disturbingly little is being done to enhance the capacity of the regulatory state to formulate and implement policy in this complex, dynamic, and divisive policy area.*[9] [our emphasis]

If multipartite bargaining is to be generally accepted as legitimate, norms, rules, and procedures will have to be developed in ways that fit into the existing political institutions and lead to a more effective regulatory environment.

Multi-Stakeholder Approaches and Democracy

These new processes may simply be vehicles for state legitimation or they may lead to genuine administrative reform (however defined). Success will depend on the extent to which public consultation penetrates the policy-making process and the quality of the ideas that emerge from the recommendations. This means that once consultation takes place, the suggestions of the roundtables would have to lead to legislation, policy formulation, regulation, and implementation, which ultimately reflects the spirit and recommendations of those initiatives. If these groups do have impact, it means that governments are listening, but are they listening to the appropriate representatives of a general public good?

This shifting policy environment may offer promise of 'more inclusive' decision-making, but it does so at a cost to the traditional representative institutions, and serves to undermine the brokerage function of the party system and the role of duly elected representatives. The nature of the trade-off is unclear. As Michael Atkinson has pointed out in a somewhat different context, 'Canadians have a different democracy than the one they inherited from their British forebears, one with its own capacity to generate stalemate and disappointment.'[10]

It has been argued that the implications for democracy do not lie so much with the nature of participation by the general public but with the role of various elites – the articulate leaders of diverse interests. Eva Etzioni-Halevy argues that democracy is best protected through the competition and challenges posed by autonomous elites:

> The elites of whatever movements for greater democracy and equality emerge in the future may well be co-opted, as their predecessors have been in the past. They may betray their rank-and-file members by becoming staunch supporters of the status quo. But as in the past, so in the future, it is precisely through their betrayal – through the price they exact for it – that changes toward greater equality may occur. And the greater the autonomy of the counter-elites at the time of co-optation, the higher the price – in concessions for their movements – which they can exact from established elites in return for their willingness to be co-opted. Hence also the greater the changes toward equality they can achieve. And if, or when, they get co-opted, others may well come to take their place.
>
> Thus the relative autonomy of elites of social movements, although it brings about only limited achievements at any given time, still turns democracy into a more dynamic and progressive system than would otherwise be the case. Not all social movements are necessarily progressive and not all necessarily deserve support. But, in general, the ability of social movements and their elites and sub-elites to express themselves autonomously (as long as they do not turn to violence) certainly needs to be defended and encouraged. And understanding the democratic role of those who instigate, organize and lead autonomous movements helps not only in understanding democracy, but also in understanding how democracy can be encouraged to breed a more democratic, equitable and egalitarian democracy in the future.[11]

If this is the case, so-called 'shared decision-making' processes will provide additional avenues for maintaining the necessary levels of accountability and responses of decision-makers to those autonomous elites.

The success of the WMI and other consultative initiatives hinges on the ability of the participants to alter their own perceptions of self-interests. For the process not to be stymied by stalemate or simply watered down to include all perspectives, participants must be able to incorporate their views into a vision broader then their own immediate personal interest or cause. They need to develop a sense of how to promote their concerns and interests through a broader understanding of the public interest.

There will always be individuals who tend to work on the margins of these consultative exercises and whose interests will never be accommodated. Although they do not represent the majority opinion, these individuals can be quite influential in shaping public opinion. In a democratic society, such activists play an important role in terms of challenging conventional wisdom and helping to create new visions of the way in which an issue could be considered. Healthy liberal democratic states require new ideas that exist on the peripheries of the mainstream conscience, as well as a political system that encourages independent thinking. On the other hand, such individuals are unlikely to work well within the confines of consensus-based processes such as the WMI. One of the challenges posed to those committed to building on the spirit of the WMI, then, is to judge how much effort should be placed on attempting to include these individuals in public consultation initiatives that require compromise; this is a process that they might not embrace in good faith.

In the roundtable exercises throughout Canada, government officials have chosen people who purport to represent a particular vested interest – and who feel strongly about that interest. How is that particular interest to be reconciled with a broader responsibility to represent the larger society? Is there any other way to hold participants accountable other than through a pledge that they will do their best to act in good faith? That role has traditionally been filled by elected representatives. These are all questions facing the Whitehorse Mining Initiative.

It is premature to come to any hard and fast conclusions about its lasting legacy or about how questions of representativeness and legitimacy will be resolved. As discussed above, however, it has considerable potential; its goals and principles may be applied in the community, provincial, national, and international arenas. The WMI provides a potential policy framework that could foster responsible and sustainable mining in the years ahead.

The creators of the Whitehorse Mining Initiative recognized that a new policy process would be required to deal with the complexity of the contemporary state if the mineral industry is to remain viable. The recommendations and goals laid out in the final accord and the reports of the issue groups will help serve as a guide for future policy-making.

Those who are ready to abandon the process as too ambitious need to be reminded that the WMI Accord demonstrates that people who hold quite different world views can arrive at a consensus. Many of the participants came to the conclusion, perhaps for the first time, that a measure of understanding and shared perspectives could be achieved. The signing of the accord was no small feat in this complex world of clashing values and competing interests. Enmeshed in seemingly endless public consultative exercises, it is not surprising that some unwilling participants would like to return to a simpler political world; one that favours their interests. As recent history has demonstrated, however, this approach would be rejected as illegitimate by the majority of Canadians who have come to expect an open and inclusive public policy process. Embracing the spirit of the Whitehorse Mining Initiative is the first step for those who hope to stake a claim in the future.

Appendix A: Chronology of Events at the WMI

22 September 1992
One-page proposal for a consensus-based initiative submitted by George Miller, president of the Mining Association of Canada, to the 49th Annual Mines Ministers' Conference in Whitehorse, Yukon.

Fall 1992
Initial planning committee, composed of sixteen to twenty members and referred to as the working group, meets in Ottawa to consider scope, objectives, and process that should define the WMI. The group will meet approximately once per month. Invitations are sent to various stakeholders to participate.

10-12 February 1993
A first meeting of the 'National Working Group' with representatives from the stakeholder groups, excluding the Assembly of First Nations and the Native Council of Canada. Decision is made to keep the process to a time-frame of twelve to fifteen months, concluding no later than September 1994. Initial budget considerations addressed.

March 1993
A two-person WMI secretariat is established. Meeting of stakeholders coincides with conference of the Prospectors and Developers Association of Canada. Further discussion about budget issues. A forty-member leadership council is announced, representing all the stakeholders.

April-June 1993
The working group meets to plan, identify issues, and define the broad mandate for issue groups. Four major themes are identified: environmental, financial and taxation, land-use, and workplace/workforce issues. Stakeholders needing financial assistance are asked to submit budget proposals.

Summer 1993-Spring 1994
Issue groups begin to meet approximately five to eight times over the next eight to ten months, for two to three days per meeting. The WMI secretariat facilitates communication and information flows.

September 1993
Annual Meeting of the Mines Ministers, New Brunswick; ministers pressure for a 'product.' Working group has sense of mission; some transition in membership. Budget shortfall is identified.

December 1993
A cross-cutting body, the Communications and Implementation Committee, is formed, with membership from all three levels – the leadership council, the working group, and the issue groups.

March 1994
The leadership council meets for its first meeting, coinciding with a meeting of the Prospectors and Developers Association. Summaries of the issue groups' work are presented to the leadership council.

May 1994
The second meeting of the leadership council, with approximately forty-four people around the table for two days. Briefing books containing issue groups' work, edited by the WMI secretariat, are pored over line by line, word by word.

May-June 1994
The Communications and Implementation Committee works to prepare a draft accord for the leadership council's consideration.

July 1994
The leadership council meets to assess and revise the draft accord. A sub-committee is formed to deal with the major unresolved issues.

July-September 1994
Consensus-making efforts, facilitated by Dan Johnston.

13 September 1994
The Whitehorse Mining Initiative Accord is submitted to the Annual Meeting of Mines Ministers, Victoria, BC.

15 October 1994
The WMI secretariat closes.

23 November 1995
A post-accord follow-up meeting of WMI stakeholders is held to assess progress.

Appendix B: WMI Vision Statement

Our vision is of a socially, economically and environmentally sustainable, and prosperous mining industry, underpinned by political and community consensus.

Mining is an important contributor to Canada's well-being, both nationally and regionally. The Whitehorse Mining Initiative is based on a shared desire to ensure that mining continues to make an important contribution within the context of sustainable development.

This vision is more simply stated than achieved. We recognize that the natural environment, the economy, and Canada's many cultures and ways of life are complex and fragile, and that each is critical to societal survival. Furthermore, no aspect of social, economic, and environmental sustainability can be pursued in isolation or be the subject of an exclusive focus without detrimentally affecting other aspects.

We also recognize that this vision will serve us well in responding to the uncertainties of the future. The context within which we seek to achieve our vision is dynamic. Social, economic, and environmental systems are constantly changing. Therefore, it is essential in realizing this vision that we enhance our ability to recognize, anticipate, and respond to change while striving to achieve a level of predictability that will allow us to pursue environmental, social, and economic goals.

The realization of this vision is not, and cannot be the responsibility of any one group. None of the stakeholders can achieve its objectives without the cooperation and support of the others. We are all aware of the need to speak plainly about the issues that face us, to think creatively about possible responses to them, and to work cooperatively to ensure that they are addressed effectively.

The Principles and Goals that we have adopted represent a major and historic first step toward revitalizing mining in Canada. They point to changes that can restore the industry's ability to attract investment for exploration and development and, at the same time, ensure that the goals of Aboriginal peoples, the environmental community, labour, and governments will be met.

The process by which we reached consensus also establishes a framework for creative cooperation, which is most important in this era of dynamic change. It is a framework that can help us anticipate, react, and adapt to changes quickly and effectively by allowing us to capitalize on the goodwill and the ability we have developed to work together, by enabling us to draw on the collective expertise of all stakeholders, and by encouraging us to resolve differences in a constructive spirit.

Source: *Searching for Gold: The Whitehorse Mining Initiative*, 14 October 1994.

Appendix C: Commitment Statement of the BC Advisory Council on Mining

Our Commitment to a Healthy, Sustainable, and Environmentally Responsible Mining Industry

Mining is woven into the fabric of British Columbia's history, society, and economy. Many communities throughout the province can trace their roots to early mining activities and many continue to depend upon mining for their prosperity. Coal mining on Vancouver Island began in 1836. Gold was discovered in the Queen Charlotte Islands in 1850, followed by the Fraser and Cariboo gold rushes. Mining has since expanded to include other metals such as copper, silver, molybdenum, lead, and zinc and is now the second largest resource industry in the province.

Mining does, however, face many challenges as it prepares to address the changing economic, environmental, and social conditions that will exist in the 21st century. To respond to these challenges, senior representatives of the mining industry, senior governments, First Nations, the environmental community, and labour from across Canada began working together in 1992 to develop a new strategic vision for mining in Canada.

In 1994, the cooperative effort resulted in the Whitehorse Mining Initiative Accord (WMI), a national accord dealing with a broad range of issues affecting or affected by mining. The WMI Accord represents a strategic vision for a healthy mining industry, in the context of maintaining healthy and diverse ecosystems, and for sharing opportunities with First Nations.

Our Vision

The British Columbia Advisory Council on Mining has been formed to further the spirit of cooperation which led to the Whitehorse Mining Initiative Accord and to proactively and cooperatively implement the accord in British Columbia.

Our vision is of a socially, economically and environmentally sustainable, accountable, and prosperous mining industry in British Columbia, underpinned by political and community support.

Our approach is based upon our recognition that the natural environment, the economy, and British Columbia's many cultures and ways of life are complex and fragile, and each is critical to the survival of a modern society. Furthermore, no aspect of social, economic, and environmental sustainability can be pursued in isolation or be the subject of an exclusive focus without detrimentally affecting other aspects. These goals must be pursued in a manner that is flexible enough to accommodate changing economic, environmental, and social requirements.

We, the British Columbia Advisory Council on Mining, are committed to our approach and are committed to achieving our vision.

Our work is based on the following understanding and principles:

- Mineral exploration and development activities must be undertaken in a manner that is responsive to the environmental, social and economic needs of local First Nations communities. First Nations should be consulted on advanced exploration activity and be involved in front end planning for new mine developments.
- The involvement, training and education of First Nations in mineral exploration and development activities is important to ensure that they receive some tangible benefits from local mining activities. Such benefits should be realized in a way that will enhance the ability of First Nations to independently engage in mineral-related development activities.

Community Needs Should Be Addressed

- Planning for new mine developments that significantly impact an existing community, or create a new community, should be undertaken in a manner that encourages long term community stability and diversity.
- Existing communities that are highly dependent on mining should seek to ensure long term stability by immediately involving all stakeholders in the development and implementation of an economic diversification plan.
- Where a mine is closed or downsized, all stakeholders should work together to develop a transition strategy for the affected workers and community.

Increased Public Involvement and Awareness Is Important

- The constructive involvement and cooperation of stakeholders is a fundamental prerequisite to generating the economic employment and social benefits available to the people of British Columbia from responsible mineral exploration and development.
- The future of a sustainable mining industry in British Columbia depends on all stakeholders working together to develop a broader public awareness and acknowledgement of mineral exploration and mining in terms of its socio-economic benefit and environmental impact.

Notes

Acknowledgments

1 The ministry has since been reorganized and is now part of the Ministry of Employment and Investment.

Introduction

1 'The Whitehorse Mining Initiative Vision,' *Searching for Gold: The Whitehorse Mining Initiative,* 14 October 1994, 11.

Chapter 1: Surveying the Terrain

1 Mining Association of Canada, 'The Mining Industry Proposes a "Whitehorse Charter,"' *Communique* (Whitehorse), 21 September 1992.
2 `The "Whitehorse Mining Initiative": A Process to Renew Canada's Minerals and Metals Sector.' Summary of recommendations from multi-stakeholder meeting, Delta Chelsea Inn, Toronto, 10-12, 26 February 1993.
3 George Hoberg, 'Environmental Policy: Alternative Styles,' *Governing Canada: Institutions and Public Policy,* ed. Michael M. Atkinson (Toronto: Harcourt Brace 1993), 318.
4 In theory, roundtables refer to a method of democratically structured, consensus-based processes where all the participants have an equal right to participate and be heard.
5 Alan Cairns and Cynthia Williams, eds., *Constitutionalism, Citizenship and Society in Canada* (Toronto: University of Toronto Press 1985).
6 Hoberg, 'Environmental Policy,' 318.
7 Mary Louise McAllister, *Prospects for the Mineral Industry: Exploring Public Perceptions and Developing Political Agendas,* Working Paper no. 50, Kingston, ON, Centre for Resource Studies, Queen's University, October 1992.
8 Correspondence with a senior government official, 12 December 1995.
9 Marsha A. Chandler, 'The Politics of Provincial Resource Policy,' *The Politics of Canadian Public Policy,* ed. Michael M. Atkinson and Marsha A. Chandler (Toronto: University of Toronto Press 1983), 44.
10 S.D. Clark, *The Developing Canadian Community,* 2nd ed. (Toronto: University of Toronto Press 1968), 248.
11 H.V. Nelles, *The Politics of Development: Forests, Mines, and Hydro-Electric Power in Ontario 1849-1941* (Toronto: Macmillan 1974), 436.
12 E.P. Herring, 'Public Administration and the Public Interest,' McGraw-Hill 1936. Reprinted with permission in Jay M. Shafritz and Albert C. Hyde, *Classics of Public Administration,* 3rd ed. (Belmont, CA: Wadsworth 1992), 76.
13 Edwin R. Black, 'British Columbia: The Politics of Exploitation,' *Party Politics in Canada,* 3rd ed., ed. Hugh G. Thorburn (Scarborough, ON: Prentice-Hall 1972), 231.
14 Alan Ewart, 'Wildland Resource Values: A Struggle for Balance,' *Society and Natural Resources* 3 (Oct.-Dec. 1990).

15 C. George Miller, president, Mining Association of Canada, 'Social Engineering in a New Era: Lessons from the Whitehorse Mining Initiative.' Presented to Resources Management in the 90s: Shared Decision Making at Work, University of Northern British Columbia, Prince George, BC, 19 October 1994.
16 John Dryzek, *Discursive Democracy: Politics, Policy and Political Science* (New York: Cambridge University Press 1990), 12.
17 Evert A. Lindquist, 'Citizens, Experts and Budgets: Evaluating Ottawa's Emerging Budget Process,' *How Ottawa Spends 1994-1995* (Ottawa: Carleton University Press 1994), 93.
18 Geraldine DeSanctis, 'Confronting Environmental Dilemmas through Group Decision Support Systems,' *The Environmental Professional* 15 (1993): 207.
19 Herring, 'Public Administration,' 77.
20 Ibid., 78.
21 W.T. Stanbury, 'A Sceptic's Guide to the Claims of So-Called Public Interest Groups,' *Canadian Public Administration* 36, no. 4 (Winter 1993): 597.
22 William D. Coleman, 'One Step Ahead: Business in the Policy Process in Canada,' *Cross-currents: Contemporary Political Issues,* 2nd ed., eds. Mark Charlton and Paul Barker (Scarborough, ON: Nelson Canada 1994), 340.
23 Stanbury, 'A Sceptic's Guide,' 581, 590.
24 T.C. Pocklington, *Liberal Democracy in Canada and the United States: An Introduction to Politics and Government* (Toronto: Holt, Rinehart and Winston 1985), 18.
25 Walter L. White, Ronald H. Wagenberg, and Ralph C. Nelson, *Introduction to Canadian Politics and Government* (Toronto: Harcourt Brace 1994), 12.
26 Patricia Marchak, Neil Guppy, and John McMullan, eds., *Uncommon Property: The Fishing and Fish-Processing Industries in British Columbia* (Toronto: Methuen 1987), 14.
27 Douglas Baker, 'Philosophical Justifications of Property Rights for Resource Allocation,' Working Paper Series no. 29, School of Urban and Regional Planning, University of Waterloo, 21.
28 Ibid., 10-21.
29 Marchak et al., *Uncommon Property,* 4.
30 Baker, 'Philosophical Justifications,' 20-1.
31 William Ophuls and A. Stephen Boyan, Jr., *Ecology and the Politics of Scarcity Revisited* (New York: W.H. Freeman 1992), 241.
32 Mary Louise McAllister, 'Local Environmental Politics,' *Metropolitics,* ed. James Lightbody (Toronto: Copp, Clark, Longman 1995), 269.
33 Susan D. Phillips, 'Whose Democratic Potential?' Comments on Liora Salter's Paper, *Rethinking Government: Reform or Reinvention?,* ed. F. Leslie Seidle (Montreal: Institute for Research on Public Policy 1993), 167.
34 Christopher Dunn, *Canadian Political Debates* (Toronto: McClelland and Stewart 1995), 199.
35 Edward Wenk, Jr., *Tradeoffs: Imperatives of Choice in a High-Tech World* (Baltimore: John Hopkins University Press 1989), 228.
36 United Nations, *Our Common Future: Report of the World Commission on the Economy and Development* (Oxford: Oxford University Press 1987), 65.
37 Donald J. Savoie, *Thatcher, Reagan, Mulroney: In Search of a New Bureaucracy* (Toronto: University of Toronto Press 1994), 184.
38 Liora Salter, 'Experiencing a Sea Change in the Democratic Potential of Regulation,' *Rethinking Government: Reform or Reinvention?,* ed. F. Leslie Seidle (Montreal: Institute for Research on Public Policy 1993), 138.
39 Phillips, 'Democratic Potential,' 162.
40 Ibid., 164.

Chapter 2: Assessing the Situation

1 Part of this chapter is based on a previous publication by Mary Louise McAllister entitled *Prospects for the Mineral Industry: Exploring Public Perceptions and Developing Political Agendas,* Working Paper no. 50, Kingston, ON, Centre for Resource Studies, Queen's University, October 1992.

2 Philip M. Hocker, 'No Mine Is an Island: The Mining Industry amid Rising Environmental Expectations,' Working Papers no. 90-95, Colorado, Mineral Policy Center, 26 April 1990.
3 Lowell Murray, then leader of the government in the Senate, Address to mineral industry representatives, *Changing Political Agendas and the Canadian Mining Industry,* ed. Mary Louise McAllister, Kingston, ON, Centre for Resource Studies, Queen's University, 1992, 1.
4 William D. Coleman, 'Canada's Mineral Industries and Interest Associations,' *Changing Political Agendas,* 57-66.
5 Ibid., 58-9.
6 Ibid., 64.
7 According to a recent New York-based government publication, *Energy New Record,* Canada was ranked third after the United States and the United Kingdom for its export of technical expertise. Natural Resources Canada, *Canada's Mining Industry: Current Situation* (Feb./Mar. 1995): 27.
8 Mining Association of Canada, *Mining in Canada: Facts and Figures - 1995* (1996), 2.
9 Ibid., 16.
10 Alan Toulin, 'Mining Coalition Presses Ottawa to Move on Regulations,' *Financial Post,* 19 October 1995, 14.
11 Alexander Ross, 'Cut, and Keep Running,' *Canadian Business* (June 1992): 102.
12 Correspondence with a senior government official, 12 December 1995.
13 Michael Porter, Harvard Business School and Monitor Company, *Canada at the Crossroads: The Reality of a New Competitive Environment.* Prepared for the Business Council on National Issues and the Government of Canada, October 1991, 99.
14 W.M. Evans, 'Canada's Space Policy,' *Tracing New Orbits: Cooperation and Competition in Global Satellite Development,* ed. Donna A. Demac (New York: Columbia University Press 1986), 131.
15 R.A. Young, Philippe Faucher, and André Blais, 'The Concept of Province-Building: A Critique,' *Perspectives on Canadian Federalism,* ed. R.D. Olling and M.W. Westmacott (Scarborough, ON: Prentice-Hall 1988), 149.
16 Government of Canada, 'Prosperity through Competitiveness,' Consultation Paper, Ottawa, Minister of Supply and Services Canada, 1991, 16.
17 Ibid.
18 'Another Step,' *Canadian Mining Journal* (February 1995): 23.
19 Harold J. Barnett and Chandler Morse, *Scarcity and Growth: The Economics of Natural Resource Availability* (Baltimore: John Hopkins Press 1963), 2.
20 Michael D. Doggett, *Incorporating Exploration in the Economic Theory of Mineral Supply* (Ph.D. diss., Department of Geological Sciences, Queen's University, 1994), 10.
21 Margot Wojciechowski, 'Canadian Minerals Industry: Strength in International Trade' (Kingston, ON: Centre for Resource Studies, Queen's University 1991).
22 Don Poirier, 'Discovery Plays Drive Market, *The Prospector* (May/June 1994): 5.
23 Brian W. Mackenzie and Michael Doggett, *Economic Potential of Mining in Manitoba: Guidelines for Taxation Policy,* Technical Paper no. 12, Kingston, ON, Centre for Resource Studies, Queen's University, April 1992, 7.
24 Blair Crawford, 'This Land Is Whose Land?' *The Northern Miner* (May 1990): 33-4.
25 'The Problem of Access to Land,' *Australian Mining Industry Council* (April 1988): 1-6.
26 David Watkins, president of Minnova, 'Back to Basics: Managing the Risks to Reap the Rewards,' Canadian Institute of Mining and Metallurgy, Mineral Economics Committee, 7th Mineral Economics Symposium, Toronto, 16 January 1992.
27 Alan Cairns, 'The Embedded State: State-Society Relations in Canada,' *State and Society: Canada in Comparative Perspective,* ed. Keith Banting (Research Coordinator), University of Toronto Press in cooperation with the Royal Commission on the Economic Union and Development Prospects for Canada and the Canadian Government (Toronto, Minister of Supply and Services Canada 1986): 55.
28 Ibid., 54.
29 Alan Cairns and Cynthia Williams, eds., *Constitutionalism, Citizenship and Society in Canada* (Toronto: University of Toronto Press 1985), 5.
30 Ibid.

31 H.V. Nelles, *The Politics of Development: Forests, Mines and Hydro-Electric Power in Ontario, 1849-1941* (Toronto: Macmillan 1974), 155.
32 Marsha A. Chandler, 'The Politics of Provincial Resource Policy,' *The Politics of Canadian Public Policy,* ed. Michael M. Atkinson and Marsha A. Chandler (Toronto: University of Toronto Press 1983), 45.
33 It should be noted, however, that as a driver of the economy, the decline of the primary sector will adversely affect the business service sector including business-related tourism, transportation, ports, dock and storage, equipment, financial and commercial business, etc.
34 Robert Presthus, *Elites in the Policy Process* (London: Cambridge University Press 1974), 7.
35 Recently, the federal government has been devoting considerable attention to the issue of harmonizing federal and provincial environmental assessment processes. The Standing Committee on Natural Resources has made recommendations for streamlining environmental regulations for mining.
36 George Hood, 'Developing Communication Strategies for Major-Project Development: The Rafferty and Alameda Dams,' *Changing Political Agendas,* Proceedings no. 25, ed. Mary Louise McAllister, Kingston, ON, Centre for Resource Studies, Queen's University, 1992, 140.
37 A. Paul Pross, *Group Politics and Public Policy* (Toronto: Oxford University Press 1986), 244.
38 Mining Association of British Columbia, 'Mining in British Columbia,' special election 1991 issue.
39 Jack Patterson, 'Multiple Use Requires Clear Leadership,' *The Northern Miner* (3 June 1991): 4.
40 Walter Segsworth, Western Resources, 'Speaking Notes,' Northern Forest Products Association Convention, 4 May 1995.
41 Ibid., 13.
42 'B.C. Land-Use Plan Sparks New Round of Confrontation,' *Globe and Mail,* 13 February 1991.
43 Brian Mackenzie and Michael Doggett, 'What the Alchemists Didn't Know: Comparing Canada's Base Metal and Gold Potential,' *CRS Perspectives* (May/June 1993): 28.
44 Justyna Laurie-Lean, 'Environmental Risks,' *Back to Basics: Managing the Risks to Reap the Rewards,* Mining Association of Canada, Canadian Institute of Mining and Metallurgy, Mineral Economics Committee, 7th Mineral Economics Symposium, Toronto, 16 January 1992.
45 'Wilderness Brawl Looms,' *Calgary Herald,* 15 October 1991, B1.
46 'Windy Craggy Retrospective: Why It Failed,' *B.C. Spaces for Nature* (24 May 1995).
47 Western Canada Wilderness Committee, 'Wild Campaign Educational Report,' co-published with *Tatshenshini Wild* 11, no. 2 (Winter/Spring 1992).
48 'Wilderness Brawl Looms.'
49 Energy, Mines and Resources Canada, 'Windy Craggy – A Development Proposal with International Transboundary Environmental Impacts,' 15.
50 The Mine Development Assessment Process requires companies to conduct environmental and socioeconomic studies before they are issued a Mine Development Certificate (MDC).
51 Price Waterhouse, *The Mining Industry in British Columbia,* June 1994, 16.
52 Crown land, as the name implies, is owned by the government. Section 109 of the 1967 *Constitution Act* vested all 'Lands, Mines, Minerals, and Royalties' in the provinces that were part of the union:

> Not only does the province originally own most natural resources, it also has regulatory authority over land and resources that have been publicly retained or privately acquired. Section 92(5) of the constitution gives the province the power to make laws in relation to exploration, development, conservation and management of non-renewable natural resources and forestry resources, including laws in relation to the rate of production from those resources. Section 92(13) gives the province further power over 'property and civil rights in the province.' This broad clause also confers on the provincial government the power to regulate the use of privately

owned land and other forms of property. Property rights are not entrenched in the constitution. In strict legal terms, therefore, the power of the province to regulate resource use, or even take back resource interests, is very broad. (Richard Schwindt, commissioner, *Report of the Commission of Inquiry into Compensation for the Taking of Resource Interests*, Province of British Columbia, Resources Compensation Commission, 21 August 1992, 6).

53 Ibid., 29.
54 *Globe and Mail,* 19 August 1995, A3.
55 'Greens Don't Want Fight, Mining Investors Told,' *Globe and Mail,* 29 May 1992.

Chapter 3: Staking a Claim

1 *The Discoverers* (Toronto: Pitt Publishing 1982), 12.
2 Ibid., 31.
3 Mining Association of British Columbia, 'Mining Prepares Political Action Plan,' *Mining Quarterly* 2, no. 1 (Spring 1995), orig. cit. in Margaret Thatcher, *The Downing Street Years* (New York: Harper Collins 1993), 167.
4 *Mining in Canada: Facts and Figures: 1995*, Mining Association of Canada in cooperation with Natural Resources Canada, 1996.
5 Joe Mavrinac, 'Kirkland Lake,' *At the End of the Shift*, ed. Matt Bray and Ashley Thomson (Toronto: Dundurn Press 1992), 148.
6 *Searching for Gold: The Whitehorse Mining Initiative*, Whitehorse Mining Initiative Leadership Council Accord, 14 October 1994, 15.
7 Manitoba Information Services, 'News Service,' 19 September 1986.
8 British Columbia, Ministry of Economic Development, Small Business and Trade, 'Natural Resource Community Fund: Guidelines,' n.d.
9 Community development is a process through which the community becomes involved in shaping its own environment in order to enhance the quality of life of its residents. Community development is an umbrella concept, involving the integration of social, economic, cultural, political, and environmental dimensions and trying to influence the processes of change in those dimensions in order to meet the community's needs and objectives more effectively. (Christopher Bryant, *Sustainable Community Development: Partnership and Winning Proposal*, Good Idea Series in Sustainable Community Development, no. 1, Sackville, NB, Rural and Small Towns Research and Studies Programme, Mount Allison University, 1991).
10 Steve Parry and Dennis Prince, 'The Save Our North Campaign,' *Changing Political Agendas*, ed. Mary Louise McAllister, Kingston, ON, Centre for Resource Studies, Queen's University 1992, 142-54.
11 'Share Prince George,' Pamphlet, 1995.
12 Jane Jacobs, *Cities and the Wealth of Nations: Principles of Economic Life* (New York: Vintage Books 1985), 224.
13 Price Waterhouse, *Breaking New Ground: Human Resource Challenges and Opportunities in the Canadian Mining Industry* (Ottawa: Minister of Supply and Services Canada 1993): 1.
14 Ibid., 26.
15 Ibid., 44.
16 Ibid., 45.
17 Richard P. Chaykowski and Anil Verma, 'Innovation in Industrial Relations: Challenges to Organization and Public Policy,' *Stabilization, Growth and Distribution: Linkages in the Knowledge Era*, ed. T.J. Courchene, 2, Bell Canada Papers on Economic and Public Policy, Kingston, ON, John Deutsch Institute, Queen's University, 1994, 368.
18 Richard P. Chaykowski and Anil Verma, 'Adjustment and Restructuring in Canadian Industrial Relations: Challenges to the Traditional System,' *Industrial Relations in Canadian Industry* (Toronto: Holt, Rinehart and Winston 1992).
19 Thomas Reid, 'Thoughts on Multi-skilling,' *Human Resource Planning for the Mining Industry*, ed. L.M. Jackson, Kingston, ON, Centre for Resource Studies, Queen's University, 1991, 47.
20 Richard P. Chaykowski, 'Industrial Relations in the Canadian Mining Industry: Transition

under Pressure,' *Industrial Relations in Canadian Industry*, 142.

21 Ken Delaney, 'Expanding Labour's Role: Co-Determination and Worker Ownership,' *Changing Political Agendas*, ed. Mary Louise McAllister, Kingston, ON: Centre for Resource Studies, Queen's University, 1992, 112.

22 The Whitehorse Mining Initiative, Workplace/Workforce/Community Issue Group, *Final Report*, November 1994.

23 Ray J. Adams and Bernard Adell, 'Unions Can Facilitate Real Change,' *Policy Options* 16, no. 8 (Oct. 1995): 16.

24 Aboriginal peoples is used here to include status Indians, Metis, and Inuit.

25 The authors are grateful to Carylin Behn of CRG projects who kindly agreed to read a draft of this section and shared her thoughtful perspectives, based on her experiences both as a band member and as a person who works for a resource company.

26 David Cole, 'Traditional Ecological Knowledge of the Naskapi and the Environmental Assessment Process,' *Law and Process in Environmental Management*, ed. Steven A. Kennett (Calgary: Canadian Institute of Resources Law 1993), 419.

27 Carylin Behn, CRG Projects, written communication, 18 February 1996.

28 Daniel Johnson, 'Northern Aboriginals, Education and Resource Development in Canada,' *CRS Perspectives*, no. 43 (March 1993): 15.

29 Ibid., 16.

30 Behn, written communication, 18 February 1996.

31 Ibid.

32 Paul Tennant, *Aboriginal Peoples and Politics* (Vancouver: University of British Columbia Press 1991), 14.

33 Frank Cassidy and Robert Bish, *Indian Investment: Its Meaning and Practice* (Lantzville, BC: Olichan Books and the Institute for Research on Public Policy), 10.

34 Ibid., 38.

35 'B.C. Land-Claim Deal Sails into Storm,' *Globe and Mail*, 16 February 1996.

36 Jackie Wolfe-Keddie, 'First Nations' Sovereignty and Land Claims: Implications for Resource Management,' *Resource and Environmental Management in Canada: Addressing Conflict and Uncertainty, 2nd ed.*, ed. Bruce Mitchell (Toronto: Oxford University Press 1995), 59.

37 Collen Getz, C.A. Walker and Associates, Management Services, 'Guidelines for Avoiding the Infringement of Aboriginal Rights: A Handbook.' Prepared for the Ministry of Energy, Mines and Petroleum Resources, British Columbia, March 1995, 8.

38 Ibid., 12.

39 Ibid.

40 Robert M. Bone, *The Geography of the Canadian North* (Toronto: Oxford University Press 1992), 226-7.

41 Robert F. Keith, 'Aboriginal Communities and Mining in Northern Canada,' *Northern Perspectives,* Canadian Arctic Resources Committee, 23, no. 3-4 (Fall/Winter 1995-6): 2-8.

42 Behn, written communication, 18 February 1996.

43 Ibid.

44 Price Waterhouse, *Breaking New Ground,* 42.

45 Behn, written communication, 18 February 1996.

46 William Ophuls and A. Stephen Boyan, Jr., *Ecology and the Politics of Scarcity Revisited* (New York: W.H. Freeman 1992), 20.

47 Mary Louise McAllister, 'Local Environmental Politics,' *Metropolitics*, ed. James Lightbody (Toronto: Copp, Clark, Longman 1995), 286.

48 Peter Berg, 'More Than Just Saving What's Left,' *Home! A Bioregional Reader* (Gabriola Island, BC: New Society Publishers 1990), 14.

49 Aldo Leopold, *The Land Ethic* (1949), 262.

50 Ibid., 263.

51 Aldo Leopold,'The Ecological Conscience,' *The River of the Mother of God,* ed. Susan L. Flader and J. Baird Callicott (Wisconsin: University of Wisconsin Press 1941), 345-6.

52 Annie Booth, 'Why I Don't Talk about Environmental Ethics Anymore,' *Trumpeter* 6, no. 4 (Fall 1989): 133.

53 Whitehorse Mining Initiative, Land Access Issue Group, *Final Report*, 1994, Appendix 6.

54 Tri-Council Statement of Commitment to Complete Canada's Networks of Protected Areas, signed in Aylmer, Quebec, 25 November 1992, cited in Whitehorse Mining Initiative, Land Access Issue Group, *Final Report*, 1994, Appendix 5.
55 McAllister, 'Local Environmental Politics.'
56 Whitehorse Mining Initiative, Finance and Taxation Issue Group, *Final Report*, November 1994, 21.
57 The MDAs were terminated in 1985.
58 For more information on government assistance see: Mary Louise McAllister and Tom F. Schneider, *Mineral Policy Update 1985-89*, Kingston, ON, Centre for Resource Studies, Queen's University, 1992.
59 Margot Wojciechowski, 'The Mineral Policy Sector: A Valuable Asset in the Federal Portfolio,' *CRS Perspectives* no. 38 (February 1992): 8.
60 Nova Scotia Department of Natural Resources, Minerals and Energy Branch, *Minerals Update* (Spring 1995), 4.
61 Whitehorse Mining Initiative, Finance and Taxation Issue Group, *Final Report*, November 1994, 21.
62 Michael E. Porter, *Canada at the Crossroads: The Reality of a New Competitive Environment*. Prepared for the Business Council on National Issues and the government of Canada, Harvard Business School and Monitor Company, October 1991, 85.
63 Bruce Mitchell, 'Beating Conflict and Uncertainty in Resource Management and Development,' *Resource Management and Development: Addressing Conflict and Uncertainty* (Toronto: Oxford University Press 1991), 272. [See also the 2nd ed. published in 1995 titled *Resource and Environmental Management in Canada: Addressing Conflict and Uncertainty*.]
64 Ibid.

Chapter 4: Rough Terrain, Rich Resource

1 David Yudelman, *Canadian Mineral Policy Formation: A Case Study of the Adversarial Process*, Working Paper no. 30, Kingston, ON, Centre for Resource Studies, Queen's University, 30 May 1984, 67.
2 Michael P. Robinson, 'Mediation Roundtables: The Recent Northwest Territories and Hawaiian Experience,' *Law and Process in Environmental Management*, ed. Steven A. Kennett (Calgary: Canadian Institute of Resources Law 1993), 368.
3 Charles K. Bens, 'Effective Citizen Involvement: How to Make It Happen,' *Civic Review* (Winter/Spring 1994): 32.
4 John Dryzek, *Discursive Democracy: Politics, Policy and Political Science* (New York: Cambridge University Press 1990), 46.
5 I. William Zartman, 'Common Elements in the Analysis of the Negotiation Process,' *Negotiation Theory and Practice*, ed. J. William Breslin and Jeffrey Z. Rubin (Cambridge, MA: Program on Negotiation at Harvard Law School 1993), 154.
6 Ibid., 14.
7 Ibid., 34.
8 Lawrence S. Bacow and Michael Wheeler, *Environmental Dispute Resolution* (New York: Plenum Press 1984).
9 Dean G. Pruitt, 'Strategic Choice in Negotiation,' *Negotiation Theory and Practice*, ed. J. William Breslin and Jeffrey Z. Rubin (Cambridge, MA: Program on Negotiation at Harvard Law School 1993), 27.
10 Ibid., 97.
11 Dan Johnston, interview, Vancouver, 21 September 1995.
12 Ibid.
13 Michael P. Robinson, 'Mediation Roundtables: The Recent Northwest Territories and Hawaiian Experience,' *Law and Process in Environmental Management*, ed. Steven A. Kennett (Calgary: Canadian Institute of Resources Law 1993), 370-1.
14 Dan Johnston, interview, 28 September 1995.
15 George Patterson, Saskatchewan Energy and Mines, Geology and Mines Branch.
16 C. George Miller, 'Social Engineering in a New Era: Lessons from the Whitehorse Mining Initiative.' Presented at a conference entitled Resources Management in the 90s: Shared

Decision-Making at Work, University of Northern British Columbia, Prince George, BC, 19 October 1994, 5.
17 Walter Segsworth, president of Westin Resources, interview, 2 November 1995.
18 George Miller, president of the Mining Association of Canada, interview, 20 October 1995.
19 Dixon Thompson, president of the Rawson Academy of Aquatic Sciences, Memorandum to the Whitehorse Mining Initiative Working Group, 27 September 1993, 2.
20 Canadian Environmental Network, *Proposal to Facilitate a Communication Plan for Canadian Environmental Network Member Participation in the 'Whitehorse Mining Initiative,'* submitted to the WMI National Working Group, 11 May 1993, 4.
21 Hans Matthews, president of the Canadian Aboriginal Minerals Association, telephone interview, 15 October 1995.
22 Hans Matthews, written correspondence, 6 February 1996.
23 Keith Conn, director of environment, Assembly of First Nations, interview, Ottawa, 6 October 1995.
24 George Patterson, Saskatchewan Energy and Mines, Geology and Mines Branch, interview.
25 Telephone interview, 14 November 1995.
26 Keith Conn, director of environment, Assembly of First Nations, interview, Ottawa, 6 October 1995.
27 George Miller, president of the Mining Association of Canada, interview, 20 October 1995.
28 Alan Young, Environmental Mining Council of British Columbia, interview, 28 September 1995.
29 Lois Hooge, head, WMI secretariat, Memorandum of the Whitehorse Mining Initiative Participant Funding Policy, 16 July 1993, 1.
30 Keith Conn, director of environment, Assembly of First Nations, interview, Ottawa, 6 October 1995.
31 Hans Matthews, president of the Canadian Aboriginal Minerals Association, telephone interview, 15 October 1995.
32 Alan Young, Environmental Mining Council of British Columbia, interview, 28 September 1995.
33 It should also be noted that during a period of numerous consultative processes, environmental groups, First Nations leaders, and others are finding it difficult to represent their interests, even if the financial resources are available. It has been observed that, 'the many demands on First Nations staff and decisionmakers at the present time should be acknowledged, including the difficulty of one individual representing the diverse interests of many bands and communities with differing goals and aspirations, limited resources, and a rapidly evolving political situation.' (BC Roundtable on the Environment and the Economy, 44.)
34 Whitehorse Mining Initiative Secretariat, Workbook for Participants in the Whitehorse Mining Initiative leadership council luncheon meeting, 14 September 1993, Section 5 – WMI Budget/Workplan, 3.
35 Alan Young, Environmental Mining Council of BC, interview, 28 September 1995.
36 Doug Hyde, environment representative, interview, October 1995.
37 Whitehorse Mining Initiative Working Group: Record of Decisions at Meeting #1, 17 February 1993, 3.
38 Ibid., 5-6.
39 C. George Miller, 'Social Engineering in a New Era,' 6.
40 Ibid.
41 Ibid.
42 Don Downe, minister of Natural Resources, interview, Halifax, NS, 15 November 1995.
43 Bert Pereboom, Peartree Consulting, interview, October 1995.
44 Doug Hyde, environment representative, interview, October 1995.

Chapter 5: The Whitehorse Mining Accord

1 *Searching for Gold: The Whitehorse Mining Initiative, A Multi-Stakeholder Approach to Renew Canada's Minerals and Metals Sector*, 9.
2 Whitehorse Mining Initiative, Finance and Taxation Issue Group, *Final Report*, November

1994, 1.

3 Whitehorse Mining Initiative, Finance and Taxation Issue Group, *Final Report,* November 1994.

4 Whitehorse Mining Initiative, Finance and Taxation Issue Group, Approved Minutes of 30 July 1993 Meeting, 1-2.

5 Finance and Taxation Issue Group, Approved Minutes, 30 July 1993, in the Whitehorse Mining Initiative Leadership Council Meeting Report, Appendices, 2-3 May 1994, 2-3.

6 Telephone interview, 15 November 1995.

7 Whitehorse Mining Initiative, Finance and Taxation Issue Group, Draft Minutes of Meeting No. 2, 13 August 1993, 2.

8 Ibid., 3.

9 Bill Toms, Chief, Resource Taxation, Tax Policy Branch, Finance Canada, interview, 20 October 1994.

10 Ibid.

11 Ibid.

12 Whitehorse Mining Initiative, Land Access Issue Group, *Final Report.*

13 Paul Dean, ADM of Energy, Mines and Resources, Newfoundland, telephone interview, 17 November 1995.

14 Interview with participant in Land Access Issue Group.

15 Whitehorse Mining Initiative, Environment Issue Group, *Final Report,* November 1994.

16 Irene Novaczek, telephone interview, 10 November 1995.

17 George Patterson, Saskatchewan Energy and Mines, Geology and Mines Branch, interview.

18 Whitehorse Mining Initiative Working Group, Outline of a proposal for a Workplace Issues Group, May 1993, 1.

19 The Whitehorse Mining Initiative, Workplace/Workforce/Community Issue Group, *Final Report,* November 1994.

20 WMI Workplace Issue Group Meeting, Third Meeting, 6-7 December 1993, Draft Minutes, 3.

21 Telephone interview, 14 November 1995.

22 Irene Novaczek, telephone interview, 10 November 1995.

23 Art Ball, interview.

24 Dan Johnston, interview.

25 WMI Land Access Issue Group, Minutes of Meeting, 27-30 November and 1 December 1993, 12.

26 Environment Issue Group, untitled document, fax transmission, 13 May 1994.

27 Ibid.

28 Issue group participant, interview.

29 Issue group member, faxed correspondence, 26 May 1994.

30 Dan Johnston, interview, 28 September 1995.

31 George Miller, president of the Mining Association of Canada, interview, 20 October 1995.

32 Interview.

33 Dixon Thompson, Memorandum to the Whitehorse Mining Initiative Working Group, 27 September 1993.

34 Dan McFadyen, ADM, Saskatchewan, telephone interview, 15 November 1995.

35 C. George Miller, 'Social Engineering in a New Era,' 6.

36 WMI Communication and Implementation ad hoc group meeting, Action Plan, 12 May 1994, 3.

37 Ibid., 2.

38 Ibid., 4.

39 Telephone and personal interviews, Fall 1995.

40 Don Downe, minister of Natural Resources, interview, Halifax, NS, 15 November 1995.

41 May meeting of the leadership council.

42 One interesting idea expressed at the May meeting was 'the need, as stakeholders, to talk to the public about what we understand about the industry.' This, unfortunately, was not carried out.

43 George Miller, president of the Mining Association of Canada, interview, 20 October 1995.

44 *Searching for Gold: The Whitehorse Mining Initiative*, 14 October 1994, 10.
45 Ibid., 25.
46 Ibid., 34.

Chapter 6: Implementing the Vision?

1 Canada, 'Federal Response,' *Whitehorse Mining Initiative Progress Reports: Implementation of the Whitehorse Mining Initiative Accord*, November 1995.
2 Ibid.
3 Bill Toms, chief, Resource Taxation, Tax Policy Branch, Finance Canada, interview, 20 October 1994.
4 Whitehorse Mining Initiative, *Federal Government Progress Report*, 'Progress Reports: Implementation of the Whitehorse Mining Initiative Accord,' November 1995.
5 Bill Toms, Chief, Resource Taxation, Tax Policy Branch, Finance Canada, interview, 20 October 1994.
6 *Federal Government Progress Report*, November 1995.
7 Ibid.
8 Ibid.
9 Ibid.
10 Department of Indian Affairs and Northern Development, 'Progress Reports: Implementation of the Whitehorse Mining Initiative Accord,' November 1995.
11 Ibid.
12 Ibid.
13 Federal Department of Fisheries and Oceans, 'Progress Reports: Implementation of the Whitehorse Mining Initiative Accord,' November 1995.
14 Darlene Smith, Department of Fisheries and Oceans, interview, October 1995.
15 Dan Adamson, Interagency Management Committee coordinator for the Omineca-Peace Land Use Resource Management Plans, interview, University of Northern British Columbia, Prince George, BC, 20 September 1995.
16 Owen, 'Participation and Sustainability,' 335.
17 Responsibilities for mineral policy are now housed in the Ministry of Employment and Investment.
18 British Columbia Advisory Council on Mining,'Terms of Reference,' 27 January 1995.
19 British Columbia Advisory Council on Mining, 'Statement of Commitment,' 2 June 1995.
20 'Overview of Some Selected Issues from Labour for Discussion by the British Columbia Advisory Council on Mining,' 19 September 1995.
21 Alan Young, Environmental Mining Council of BC, 'Key Issues in Mine Development and Land-Use Planning: An Environmental Perspective for Discussion by the Minister's Advisory Council on Mining Submitted to the BC Advisory Council on Mining,' September 1995.
22 Discussion paper prepared for the British Columbia Advisory Council on Mining by the Mining Association of British Columbia, the Mining Suppliers, Consultants and Contractors Association, and the BC and Yukon Chamber of Mines, 8 September 1995.
23 Ibid.
24 Canada/British Columbia, 'Canada and British Columbia to Cooperate on Environmental Assessment,' News Release, 23 October 1995.
25 Ministry of Environment, Lands and Parks, Ministry of Energy, Mines and Petroleum Resources, 'Skagit Valley to Become a Provincial Park,' News Release, 21 September 1995.
26 Ibid.
27 Reclamation Security Policy Task Force Committee, *Report and Recommendations to the Minister of Energy, Mines and Petroleum Resources on British Columbia Mining Reclamation Security Policies*, version 3, 6 February 1996, 5.
28 Ibid., 30.
29 Ibid., 67.
30 Ibid., 68.
31 BC Energy, Mines and Petroleum Resources, BC Advisory Council on Mining, 'Re: Special Management Zone Workshop, Summary Notes and Recommendations,' 13 December 1995.

32 Anne Edwards, minister of mines, BC, interview, 1 November 1995.
33 Annie Booth, 'Why I Don't Talk about Environmental Ethics Anymore,' *The Trumpeter: Journal of Ecosophy* 6, no. 4 (Fall 1989): 133.
34 Ontario, Commission on Planning and Development Reform in Ontario, *Draft Report,* 18 December 1992, 1.
35 Ibid., 136-42.
36 Ibid.
37 Ontario, Ministry of Municipal Affairs, 'Comprehensive Set of Policy Statements' (Toronto: Queen's Printer 1994), 16-17.
38 'Ontario's Environmental Image a Tough Sell,' *Globe and Mail*, 4 November 1996, A3.
39 Ontario, Ministry of Northern Development and Mines, 'Ontario Has Implemented Major WMI Recommendations Dealing with Mine Reclamation and Land-Use Planning,' September 1994.
40 Ontario, Ministry of Natural Resources, 'Response to the Whitehorse Mining Initiative,' July 1995.
41 Ibid.
42 Ontario, Ministry of Northern Development and Mines, correspondence, 26 January 1996.
43 Ontario, Ministry of Northern Development and Mines, 'Background: Mining Act Amendments Reduce Unnecessary Costs and Red Tape,' 1 November 1995.
44 Canadian Environmental Law Association, 'Cutting Ontario's Environment,' Special Report, April 1996, 6.
45 *Globe and Mail*, 18 October 1996, A9.
46 New Brunswick, 'Whitehorse Mining Initiative: Progress Reports.'
47 Don Barnett, ADM, New Brunswick, Department of Minerals and Energy, telephone interview, 17 November 1995.
48 Ibid.
49 Nova Scotia, 'Whitehorse Mining Initiative: Progress Reports,' 1.
50 Pat Phelan, executive director, Minerals and Energy Branch, Nova Scotia, Department of Natural Resources, interview, 15 November 1995.
51 Don Downe, minister of Natural Resources, Halifax, NS, interview, 15 November 1995.
52 Newfoundland and Labrador, 'Whitehorse Mining Initiative: Progress Reports.'
53 Paul Dean, ADM, Newfoundland, Mineral Resources, telephone interview, 17 November 1995.
54 Ibid.
55 Newfoundland and Labrador, 'WMI Progress Reports: Implementation of the Whitehorse Mining Initiative Accord,' November 1995, 1-3.
56 Paul Dean, ADM, Newfoundland, Mineral Resources, telephone interview, 17 November 1995.
57 Input on WMI Accord implementation by the Inuit Tapirisat of Canada, addition to 'Progress Reports: Implementation of the Whitehorse Mining Initiative Accord,' November 1995, 2.
58 Paul Dean, ADM, Newfoundland, Mineral Resources, telephone interview, 17 November 1995.
59 Ibid.
60 Saskatchewan, *The Saskatchewan Mining Task Force Report*, September 1994, title page.
61 Ibid., 6.
62 Dan McFadyen, ADM, Saskatchewan, Energy and Mines, telephone interview, 15 November 1995.
63 Ibid.
64 In this context, the area is protected from all other uses.
65 George Patterson, Saskatchewan Department of Energy and Mines, Geology and Mines Branch, telephone interview, 17 November 1995.
66 Saskatchewan, Saskatchewan Department of Energy and Mines, *Action Plan,* faxed transmission.
67 Dan McFadyen, ADM, Saskatchewan, Energy and Mines, telephone interview, 15 November 1995.

68 Ibid.
69 Ed Huebert, VP, Mining Association of Manitoba, telephone interview, 16 November 1995.
70 Ibid.
71 Ibid.
72 Ibid.
73 Ibid.
74 Ibid.
75 Ed Huebert, VP, Mining Association of Manitoba, telephone interview, 17 November 1995.
76 Manitoba, 'Whitehorse Mining Initiative: Progress Reports,' 4.

Chapter 7: Perspectives on the Accord

1 *Searching for Gold: The Whitehorse Mining Initiative, A Multi-Stakeholders Approach to Renew Canada's Minerals and Metals Sector,* 14 October 1994, 165.
2 Written communications, 6 February 1996.
3 Written comments, 27 November 1995.
4 Inuit Tapirisat of Canada, 'Input on WMI Accord Implementation by the Inuit Tapirisat of Canada, WMI follow-up meeting, Ottawa, 23 November 1995.
5 *Searching for Gold,* 166.
6 Interview, November 1995.
7 Interview, November 1995.
8 *Searching for Gold,* 170.
9 Ibid., 167.
10 United Steelworkers of America, 'Status Report for WMI Follow-Up Meeting,' *Whitehorse Mining Initiative Progress Reports: Implementation of the Whitehorse Mining Initiative Accord,* November 1995.
11 Telephone interview.
12 *Searching for Gold,* 168.
13 Telephone interview, 31 October 1995.
14 Telephone interview.
15 Interview, 1 November 1995.
16 *Searching for Gold,* 169.
17 Bob Van Dijken, Yukon Conservation Society, written notes, 28 November 1995.
18 Interview.
19 Interview.
20 'The Best Political Decision Money Can Buy,' *Financial Post,* 11-12.
21 'The Nasty Game,' *Alternatives Journal* (Oct.-Nov. 1996): 10-17.
22 'Canada's First Diamond Mine Approved,' *Toronto Star,* 2 November 1996.

Chapter 8: Terra Incognita

1 David M. Cameron, *Canadian Public Administration* 37, no. 3 (Fall 1994): 399.
2 *The Northern Miner,* 19 April 1993. (Also interview with planner on 15 November 1995.)
3 Arthur Ball, director of mines, Manitoba Department of Energy and Mineral Resources, telephone interview, 17 November 1995.
4 *Canadian Mining Journal* (Aug. 1994): 15.
5 'A Future Denied,' Supplement, *Report on Business,* September 1995.
6 Whitehorse Mining Initiative, *Leadership Council Accord: Final Report,* November 1994, 29.
7 Evert A. Lindquist, 'Citizens, Experts and Budgets: Evaluating Ottawa's Emerging Budget Process,' *How Ottawa Spends 1994-95* (Ottawa: Carleton University Press 1994), 428.
8 Hoberg, 'Environmental Policy,' 338.
9 Ibid., 339.
10 Michael Atkinson, 'What Kind of Democracy Do Canadians Want?', *Canadian Journal of Political Science* 27, no. 4 (Dec. 1994): 745.
11 Eva Etzioni-Halevy, *The Elite Connection: Problems and Potential of Western Democracy* (Cambridge, MA: Polity Press 1993): 213-14.

Select Bibliography

Books and Articles

Alder, Nancy J. *International Dimensions of Organizational Behavior.* Boston: Kent Publishing Company 1986

Australian Mining Industry Council. *The Problem of Access to Land.* April 1988

Bacow, Lawrence S., and Michael Wheeler. *Environmental Dispute Resolution.* New York: Plenum Press 1984

Baker, Douglas. *Philosophical Justifications of Property Rights for Resource Allocation.* Working Paper Series no. 29. Waterloo: School of Urban and Regional Planning, University of Waterloo, n.d.

Barnaby, George, band councillor, Fort Good Hope. 'The Goal of Economic Development.' *Community Economic Development in Canada's North.* Ottawa: Canadian Arctic Resource Committee 1990

Barnett, Harold J., and Chandler Morse. *Scarcity and Growth: The Economics of Natural Resource Availability.* Baltimore: John Hopkins Press 1963

Bauer, Richard J. 'A Customer Perspective.' *Marketing of Nonferrous Metals: Proceedings of the Nineteenth CRS Policy Discussion Seminar.* 21-3 Nov. 1988. Ed. L. Moira Jackson and Peter R. Richardson. Kingston, ON: Centre for Resources Studies August 1989

Black, Edwin, R. 'British Columbia: The Politics of Exploitation.' *Party Politics,* 3rd ed. Ed. Hugh G. Thorburn. Scarborough, ON: Prentice-Hall 1972

Bloom, Allen. *Closing of the American Mind.* New York: Simon and Schuster 1987

Bone, Robert M. *The Geography of the Canadian North.* Toronto: Oxford University Press 1992

Bray, Matt, and Ashley Thomson. *At the End of the Shift: Mines and Single Industry Towns in Northern Ontario.* Toronto and Oxford: Dundurn Press 1992

Brooks, Stephen, and Andrew Stritch. *Business and Government in Canada.* Scarborough, ON: Prentice-Hall 1991

Bryant, Christopher. *Sustainable Community Development, Partnerships and Preparing Winning Proposals.* Good Idea Series in Sustainable Community Development, no. 1, Rural and Small Towns Research and Studies Programme. Eugne, ON: Mount Allison University, NB, and Ecotrends 1991

Brown, Peter G. *Restoring the Public Trust: Fresh Vision for Progressive Government in America.* Boston: Beacon Press 1994

Buck, W. Keith. *Intergovernmental Mineral Commodity Arrangements.* Working Paper no. 37. Kingston, ON: Centre for Resource Studies, Queen's University September 1986

Cairns, Alan. 'The Embedded State: State-Society Relations in Canada.' *State and Society: Canada in Comparative Perspective.* Ed. Keith Banting (Research Coordinator), University of Toronto Press in cooperation with the Royal Commission on the Economic Union and Development Prospects for Canada and the Canadian Government. Toronto: Minister of Supply and Services Canada 1986

Cairns, Alan, and Cynthia Williams, eds. *Constitutionalism, Citizenship and Society in Canada*. Toronto: University of Toronto Press 1985

Cassidy, Frank. 'Aboriginal Land Claims in British Columbia.' *Aboriginal Land Claims in Canada*. Ed. Ken Coates. Toronto: Copp, Clark, Pitman 1992

CAW Canada. 'CAW Is Working People Like You.' Pamphlet, n.d.

Chandler, Marsha A. 'The Politics of Provincial Resource Policy.' *The Politics of Canadian Public Policy*. Ed. Michael M. Atkinson and Marsha A. Chandler. Toronto: University of Toronto Press 1983

Chaykowski, Richard P., and Anil Verma. 'Innovation in Industrial Relations: Challenges to Organizations and Public Policy.' *Stabilization, Growth and Distribution: Linkages in the Knowledge Era*. Bell Canada Papers on Economic and Public Policy, 2, ed. T.J. Courchene. Kingston, ON: John Deutsch Institute, Queen's University 1994

–. 'Adjustments and Restructuring in Canadian Industrial Relations: Challenges to the Traditional System.' *Industrial Relations in Canadian Industry*. Toronto: Holt, Rinehart and Winston 1992

Clark, S.D. *The Developing Canadian Community*. 2nd ed. Toronto: University of Toronto Press 1968

Coates, Ken, ed. *Aboriginal Land Claims in Canada*. Toronto: Copp, Clark, Pitman 1992

Cole, David. 'Traditional Ecological Knowledge of the Naskapi and the Environmental Assessment Process.' *Law and Process in Environmental Management*. Ed. Steven A. Kennett. Calgary: Canadian Institute of Resource Law 1993

Coleman, William D. 'One Step Ahead: Business in the Policy Process in Canada.' *Crosscurrents: Contemporary Political Issues*. 2nd ed. Ed. Mark Charlton and Paul Barker. Scarborough, ON: Nelson 1994

Dahl, Robert A. *After the Revolution? Authority in a Good Society*. New Haven: Yale University Press 1970

Davis, Albie M. 'An Interview with Mary Parker Follett.' *Negotiation Theory and Practice*. Ed. J. William Breslin and Jeffrey Z. Rubin. Cambridge, MA: Program on Negotiation at Harvard Law School 1993

DePape, Denis. 'Alternatives to Single Project Mining Communities: A Critical Assessment.' *Mining Communities: Hard Lessons for the Future*. Intergroup Consulting Economists, n.d.

The Discoverers. Toronto: Pitt Publishing 1982

Doggett, Michael D. *Incorporating Exploration in the Economic Theory of Mineral Supply*. Ph.D. diss., Queen's University March 1994

Dryzek, John. *Discursive Democracy: Politics, Policy and Political Science*. New York: Cambridge University Press 1990

Dunn, Christopher. *Canadian Political Debates*. Toronto: McClelland and Stewart 1995

Etzioni-Halevy, Eva. *The Elite Connection: Problems and Potential of Western Democracy*. Cambridge, MA: Polity Press 1993

Evans, W.M. 'Canada's Space Policy.' *Tracing New Orbits: Cooperation and Competition in Global Satellite Development*. Ed. Donna A. Demac. New York: Columbia University Press 1986

Ewart, Alan. 'Wildland Resource Values: A Struggle for Balance.' *Society and Natural Resources* 3 (Oct.- Dec. 1990): 385-93

Fraser, Bruce, Salasan Associates. *Extracting the Lessons from the CORE Land-Use Planning Process in the Kootenay-Boundary Region of British Columbia*. Mediator's Report. 27 July 1994

Getz, Colleen, C.A. Walker and Associates, Management Services. 'Guidelines for Avoiding the Infringement of Aboriginal Rights: A Handbook.' Prepared for the Ministry of Energy, Mines and Petroleum Resources, British Columbia. March 1995

Glasser, Theodore L. 'Communication and the Cultivation for Citizenship.' *Communication* 12 (1991)

Hanson, Bill. 'Native Issues in Long Distance Commuting Mining.' *Long Distance Commuting in the Mining Industry: Conference Summary*. Proceedings no. 24. Ed. Mark Shrimpton and Keith Storey, 1991

Herring, E.P. 'Public Administration and the Public Interest.' McGraw-Hill 1936. Reprinted

with permission in Jay M. Shafritz and Albert C. Hyde. *Classics of Public Administration.* 3rd ed. Belmont, CA: Wadsworth 1992
Hinde, Christopher. 'Global Economic Trends and Increases in Long Distance Community.' *Long Distance Commuting in the Mining Industry: Conference Summary*. Proceedings no. 24. Ed. Mark Shrimpton and Keith Storey. Kingston, ON: Energy, Mines and Petroleum Resources Canada, Centre for Resource Studies, Queen's University 26-28 November 1990
Hoberg, George. 'Environmental Policy: Alternative Styles.' *Governing Canada: Institutions and Public Policy*. Ed. Michael M. Atkinson. Toronto: Harcourt Brace 1993
Hocker, Philip M. 'No Mine Is an Island: The Mining Industry amid Rising Environmental Expectations.' Working Papers no. 90-5. Colorado: Mineral Policy Center, Colorado School of Mines 26 April 1990
Issac, Jeffrey C. *Ardent, Camus, and Modern Rebellion*. New Haven: Yale University Press 1992
Jackson, Moira L.'The Role of Education and Training in the Mining Industry.' Draft. Kingston, ON: Centre for Resource Studies, Queen's University 9 June 1995
–. 'The Status Quo Is Not an Option.' Paper presented at the 92nd Annual General Meeting. Ottawa 6-10 May 1990
Jacobs, Jane. *Cities and the Wealth of Nations: Principles of Economic Life*. New York: Vintage Books 1985
Kernaghan, Kenneth, and David Siegal. *Public Administration in Canada*. 2nd ed. Scarborough, ON: Nelson 1991
Laurie-Lean, Justyna. 'Environmental Risks.' *Back to Basics: Managing the Risks to Reap the Rewards*. Toronto: Mining Association of Canada, Canadian Institute of Mining and Metallurgy, Mineral Economics Committee, 7th Mineral Economics Symposium 16 January 1992
Leopold, Aldo. *The Land Ethic*. 1949
Lindquist, Evert A. 'Citizens, Experts and Budgets: Evaluating Ottawa's Emerging Budget Process.' *How Ottawa Spends 1994-95*. Ottawa: Carleton University Press 1994
Lucas, Rex A. *Minetown, Milltown, Railtown: Life in Canadian Communities of Single Industry*. Toronto: University of Toronto Press 1971
Mackenzie, Brian W., and Michael Doggett. *Economic Potential of Mining in Manitoba: Guidelines for Taxation Policy*. Technical Paper no. 12. Kingston, ON: Centre for Resource Studies, Queen's University April 1992
Mackenzie, Brian W., Michel Bilodeau, and Michael Doggett. *Mineral Exploration and Mine Development Potential in Ontario: Economic Guidelines for Government Policy*. Technical Paper no. 9 April 1989
Mansbridge, Jane. *Beyond Adversary Democracy*. Chicago: University of Chicago Press 1983
Marchak, Patricia, Neil Guppy, and John McMullan, eds. *Uncommon Property: The Fishing and Fish-Processing Industries in British Columbia*. Toronto: Methuen 1987
McAllister, Mary Louise. 'Local Environmental Politics.' *Metropolitics*. Ed. James Lightbody. Toronto: Copp, Clark, Longman 1995
–. *Prospects for the Mineral Industry: Exploring Public Policy Perceptions and Developing Political Agendas*. Working Paper no. 50. Kingston, ON: Centre for Resource Studies, Queen's University October 1992
–, ed. *Changing Political Agendas and the Canadian Mining Industry*. Kingston, ON: Centre for Resource Studies, Queen's University 1992
McAllister, Mary Louise, and Tom F. Schneider. *Mineral Policy Update 1985-89*. Kingston, ON: Centre for Resource Studies, Queen's University 1992
McGeer, Peter. 'The Challenge of New Materials.' *Prospects for Minerals in the 90s*. Proceedings no. 21. Ed. Peter R. Richardson and Renka Gesing. Kingston, ON: Centre for Resource Studies, Queen's University November 1988
McKnight, Bruce. 'Core Relations.' Presented at the 99th Northwest Mining Association Annual Convention, Spokane, WA, 28 Nov.-3 Dec. 1993
–. 'Vancouver Island – On CORE and Encore.' Kamloops Exploration Group, KEG '94 11-12 April 1994
Melucci, Alberta. 'Social Movements and the Democratization of Everyday Life.' *Civil Society and the State: New European Perspectives*. Ed. John Keans. New York: Verso 1993

Mills, Claudia, and Douglas Maclean. 'Faith in Science.' *Values and Public Policy*. Ed. Claudia Mills. Toronto: Harcourt Brace Jovanovich 1992

Mining Association of British Columbia. 'Coal Mining and the CORE Process in the East Kootenays.' *Mining Quarterly* 2, no. 1 (Spring 1995)

–. 'Mining Prepares Political Action Plan.' *Mining Quarterly* 2, no. 1 (Spring 1995)

–. *Mining Quarterly* 1, no. 2 (Fall 1994)

–. 'Newsletter: Mining in British Columbia.' Special Election Issue 1991

Mining Association of Canada, in Cooperation with Natural Resources Canada. 'Mining in Canada: Facts and Figures: 1993.' 1994

Mitchell, Bruce. 'Beating Conflict and Uncertainty in Resource Management and Development.' *Resource Management and Development: Addressing Conflict and Uncertainty*. Toronto: Oxford University Press 1991

Moon, Donald. *Constructing Community: Moral Pluralism and Tragic Conflicts*. Princeton, NJ: Princeton University Press 1993

Mouffe, Chantal. *The Return of the Political*. New York: Verso 1993

Mumford, Lewis. *The Culture of Cities*. 1938. New York: Harcourt Brace Jovanovich 1970

Nelles, H.V. *The Politics of Development: Forests, Mines, and Hydro-Electric Power in Ontario 1849-1941*. Toronto: Macmillan 1974

Ophuls, William, and A. Stephen Boyan, Jr. *Ecology and the Politics of Scarcity Revisited*. New York: W.H. Freeman 1992

Owen, Stephen. 'Participation and Sustainability: The Imperatives of Resource and Environmental Management.' *Law and Process in Environmental Management: Essays from the Sixth CIRL Conference on Natural Resource Law*. Ed. Steven A. Kennett. Calgary: Canadian Institute of Resource Law 1993

Paget, Gary, and Brian Walisser. 'The Development of Mining Communities in British Columbia: Resilience through Local Governance.' *Mining Communities: Hard Lessons for the Future*. Proceedings of the 12th CRS Policy Discussion Seminar. Kingston, ON: Centre for Resource Studies, Queen's University 27-29 September 1983

Phillips, Susan D. 'Whose Democratic Potential?' Comments on Liora Salter's Paper *Rethinking Government: Reform or Reinvention?* Ed. F. Leslie Seidle. Montreal: Institute for Research on Public Policy 1993

Pocklington, T.C. *Liberal Democracy in Canada and the United States: An Introduction to Politics and Government*. Toronto: Holt, Rinehart and Winston 1985

Porter, Michael E. *Canada at the Crossroads: The Reality of a New Competitive Environment*. A study prepared for the Business Council on National Issues and the government of Canada, Harvard Business School and Monitor Company. October 1991

Presthus, Robert. *Elites in the Policy Process*. London: Cambridge University Press 1974

Price Waterhouse. *Breaking New Ground: Human Resource Challenges and Opportunities in the Canadian Mining Industry*. Ottawa: Minister of Supply and Services Canada 1993

–. *The Mining Industry in British Columbia 1993*. June 1994

Pross, A. Paul. *Group Politics and Public Policy*. Toronto: Oxford University Press 1986

Pruitt, Dean G. 'Strategic Choice in Negotiation.' *Negotiation Theory and Practice*. Ed. J. William Breslin and Jeffrey Z. Rubin. Cambridge, MA: Program on Negotiation at Harvard Law School 1993

Reid, Thomas. 'Thoughts on Multi-skilling.' *Human Resource Planning for the Mining Industry*. Ed. L. Moira Jackson. Proceedings of the Twentieth CRS Policy Discussion Seminar, 26-28 March 1990. Kingston, ON: Centre for Resource Studies, Queen's University August 1991

Richardson, Mary, Joan Sherman, and Michael Gismondi. *Winning Back the Words: Confronting Experts in an Environmental Public Hearing*. Toronto: Garmond Press 1993

Robinson Consulting and Associates, Gary Schaan Consulting. *Land Resources Joint Management Arrangements*. Victoria March 1995

Salter, Liora. 'Experiencing a Sea Change in the Democratic Potential of Regulation.' *Rethinking Government: Reform or Reinvention?* Ed. F. Leslie Seidle. Montreal: Institute for Research on Public Policy 1993

Saven, Beth. *Science under Siege: The Myth of Objectivity in Scientific Research*. Toronto: Canadian Broadcasting Corporation 1988

Savoie, Donald J. *Thatcher, Reagan, Mulroney: In Search of a New Bureaucracy*. Toronto: University of Toronto Press 1994

Sennett, Richard. *The Fall of Public Man*. New York: W.W. Norton 1976

Segsworth, Walter, Westmin Mines. 'Speaking Notes.' Northern Forest Products Association Convention, Prince George, BC 4 May 1995

'Share Prince George.' Pamphlet, 1995

Sorrell, Charles A. 'New Materials: Technology Overview.' *The New Materials Society: Opportunities and Challenges*. Vol. 2. US Department of the Interior/Bureau of Mines 1900

Stevens, Terry. 'Long Distance Commuting, Mining and Organized Labour.' *Long Distance Commuting in the Mining Industry: Conference Summary*. Proceedings no. 24. Ed. Mark Shrimpton and Keith Storey. Kingston, ON: Centre for Resource Studies, Queen's University, Energy, Mines and Petroleum Resources Canada September 1991

Storey, Keith, and Mark Shrimpton. *Long Distance Labour Commuting in the Canadian Mining Industry*. Working Paper no. 43. Kingston, ON: Centre for Resources Studies, Queen's University 1988

Tennant, Paul. *Aboriginal Peoples and Politics*. Vancouver: University of British Columbia Press 1991

United Nations. *Our Common Future: Report of the World Commission on the Economy and Development*. Oxford: Oxford University Press 1987

Wallace, Iain. *The Global Economic System*. London: Unwin Hyman 1990

Watkins, David, president of Minnova. 'Back to Basics: Managing the Risks to Reap the Rewards.' Canadian Institute of Mining and Metallurgy, Mineral Economics Committee. 7th Mineral Economics Symposium, Toronto 16 January 1992

Wenk, Edward, Jr. *Tradeoffs: Imperatives of Choice in a High-Tech World*. Baltimore: John Hopkins University Press 1989

Western Canada Wilderness Committee. 'Wild Campaign Educational Report.' Co-published with *Tatshenshini* 11, no. 2 (Winter/Spring 1992)

Whalen, Hugh. *Public Participation and the Role of Canadian Royal Commissions and Task Forces 1957-1969*. Paper presented at the 1981 Conference of the Institute of Public Administration of Canada, Charlottetown, PEI, 8-11 September 1981

White, Walter L., Ronald H. Wagenberg, and Ralph C. Nelson. *Introduction to Canadian Politics*. Toronto: Harcourt Brace 1994

Wilson, Jeremy. 'Wilderness Politics in B.C.' *Policy Communities and Public Policy in Canada: A Structural Approach*. Ed. William D. Coleman and Grace Skogstad. Toronto: Copp, Clark, Pitman 1990

Wojciechowski, Margot. 'Canadian Minerals Industry: Strength in International Trade.' Kingston, ON: Centre for Resource Studies, Queen's University 1991

Wolfe-Keddie, Jackie. '"First Nations" Sovereignty and Land Claims: Implications for Resource Management.' *Resource and Environmental Management in Canada: Addressing Conflict and Uncertainty*. 2nd ed. Ed. Bruce Mitchell. Toronto: Oxford University Press 1995

Yudelman, David. *Canadian Mineral Policy Formulation: A Case Study of the Adversarial Process*. Working Paper no. 30. Kingston, ON: Centre for Resource Studies, Queen's University 1984

Young, Alan. 'Key Issues in Mine Development and Land Use Planning: An Environmental Perspective for Discussion by the Minister's Advisory Council on Mining Submitted to the BC Advisory Council on Mining.' Environmental Council of BC September 1995

Young, R.A., Philippe Faucher, and André Blais. 'The Concept of Province-Building: A Critique.' *Perspectives on Canadian Federalism*. Ed. R.D. Olling and W.M. Westmacott. Scarborough, ON: Prentice-Hall 1988

Zartman, I. William. 'Common Elements in the Analysis of the Negotiation Process.' *Negotiation Theory and Practice*. Cambridge, MA: Program on Negotiation at Harvard Business School 1993

Government Publications

British Columbia. Commission on Resources and Environment. *Public Participation: The Provincial Land Use Strategy*. 3, Victoria February 1995

–. Commission on Resources and Environment. *Report on a Land Use Strategy for British Columbia*. Victoria August 1992
–. Commission on Resources and Environment. *A Sustainability Act for British Columbia: Consolidating the Progress. Securing the Future*. 1, Victoria November 1994
–. Commission on Resources and Environment. *Vancouver Island Land Use Plan*. Victoria February 1994
–. Commission on Resources and Environment. *West Kootenay-Boundary Land Use Plan*. Victoria October 1994
–. *Local Round Tables: Realizing Their Full Potential*. British Columbia Round Table on the Environment and the Economy, June 1994
–. *Mineral Market Update*. Ministry of Energy, Mines and Petroleum Resources. 2, no. 1, January 1990
–. Ministry of Energy, Mines and Petroleum Resources. 'The Mining Industry in British Columbia: A Statistical Summary of Recent Trends.' Prepared for the Advisory Council on Mining. Jacob Mathew. Victoria 23 January 1995
–. *Natural Resource Community Fund: Guidelines*. Ministry of Economic Development, Small Business and Trade.
–. 'Skagit Valley to Become a Provincial Park.' News Release. Ministry of Environment, Lands and Parks, Ministry of Energy, Mines and Petroleum Resources. 21 September 1995
British Columbia Advisory Council on Mining.'Overview of Some Selected Issues for Discussions by the British Columbia Advisory Council on Mining.' 19 September 1995
–. Schwindt, Richard, Commissioner. *Report of the Commission of Inquiry into Compensation for the Taking of Resource Interests*. Victoria, Province of British Columbia, Resource Compensation Commission, 21 August 1992
–. *Statement of Commitment*. 2 June 1995
–. *Terms of Reference*. 27 January 1995
Canada. *Canada's Mining Industry: Current Situation*. Natural Resources Canada, Feb./Mar. 1995
–. 'Federal Response: Whitehorse Mining Initiative Progress Reports: Implementation of the WMI Accord.' November 1995
–. *Prosperity through Competitiveness*. Consultation Paper. Minister of Supply and Services Canada. Ottawa 1991
–. *Windy Craggy: A Development Proposal with International Transboundary Environmental Impacts*. Energy, Mines and Resources Canada. n.d.
Manitoba Information Services. 'News Service.' 19 September 1986
Nova Scotia. *Minerals Update*. Nova Scotia Department of Natural Resources, Minerals and Energy Branch, Spring 1995
Ontario. 'Background: Mining Act Amendments Reduce Unnecessary Costs and Red Tape.' Toronto: Ministry of Northern Development and Mines 1 November 1994
–. Commission on Planning and Development Reform in Ontario. *Draft Report*. Toronto 18 December 1992
–. Commission on Planning and Development Reform in Ontario. *New Planning for Ontario: Final Report*. Jon Sewell, Chair. Toronto: Publications Ontario June 1993
–. 'Ontario Has Implemented Major WMI Recommendations Dealing with Mine Reclamation and Land-Use Planning.' Backgrounder. Toronto: Ministry of Northern Development and Mines September 1994
–. *Ontario's New Planning System*. Toronto: Ministry of Municipal Affairs December 1994
–. 'Response to the Whitehorse Mining Initiative.' Toronto: Ministry of Natural Resources July 1995
–. *Understanding Ontario's Planning Reform*. Toronto: Ministry of Municipal Affairs 1994
Saskatchewan. *The Saskatchewan Mining Task Force Report*. September 1994
–. *Action Plan*. Saskatchewan Department of Energy and Mines. Faxed Transmission.

Magazines and Journals

Adams, Ray J., and Bernard Adell. 'Unions Can Facilitate Real Changes.' *Policy Options* 16, no. 8 (October 1995)

'A Future Denied.' *Report on Business*. Supplement (September 1995)

'Argentine Government Agency Woos Canadian Mining Companies.' *The Northern Miner* (28 August 1995)

Atkinson, Michael. 'What Kind of Democracy Do Canadians Want?' *Canadian Journal of Political Science* (December 1994)

Bens, Charles K. 'Effective Citizen Involvement: How to Make It Happen.' *Civic Review* (Winter/Spring 1994)

Bonnicksen, Thomas M. 'The Impact Process: A Computer-Aided Group Decision-Making Procedure for Resolving Complex Issues.' *The Environmental Professional* 15 (1993)

Booth, Annie. 'Why I Don't Talk about Environmental Ethics Anymore.' *The Trumpeter: Journal of Ecosophy* 6, no. 4 (Fall 1989)

Canadian Mining Journal. 'Another Step.' (February 1995)

–. (August 1994)

Canadian Public Administration 37, no. 3 (Fall 1994)

Crawford, Blair. 'This Land Is Whose Land?' *The Northern Miner* (May 1990)

DeSanctis, Geraldine. 'Confronting Environmental Dilemmas through Group Decision Support Systems.' *Environmental Professional* 15 (1993)

Hull, Dale L. 'Factors Affecting the Performance of the Canadian Minerals and Metals Industry.' *Raw Materials Report* 5, no. 3 (1987)

Johnson, Daniel. 'Northern Aboriginals, Education and Resource Development in Canada.' *CRS Perspectives* no. 43 (March 1993)

Keith, Robert F. 'Aboriginal Communities and Mining in Northern Canada.' *Northern Perspectives* 23, no. 3-4 (Fall/Winter 1995-6)

Patterson, Jack. 'Multiple Use Requires Clear Leadership.' *The Northern Miner* (3 June 1991)

Poirier, Don. 'Discovery Plays Drive Market.' *The Prospector* (May-June 1994)

Ross, Alexander. 'Cut, and Keep Running.' *Canadian Business* (June 1992)

Scott, David. 'Snip, Snip Hooray.' *Canadian Mining Journal* (Aug. 1994)

Stanbury, W.T. 'A Sceptic's Guide to the Claims of So-called Public Interest Groups.' *Canadian Public Administration* 36, no. 4 (Winter 1993)

'Windy Craggy Retrospective: Why It Failed.' *B.C. Spaces for Nature* (24 May 1995)

Wojciechowski, Margot. 'The Mineral Policy Sector: A Valuable Asset in the Federal Portfolio.' *CRS Perspectives* no. 38 (February 1992)

Newspaper Sources

'B.C. Land-Use Plan Sparks New Round of Confrontation.' *Globe and Mail* (13 February 1991)

'B.C. Sets Up Land Study.' *Globe and Mail* (22 July 1992)

'Cariboo Land-Use Plan Lambasted.' *Province* (15 July 1994)

'Core Forgets: Jobs Are People.' *Times-Colonist* (15 July 1994)

'Electric Cars to Spur Nickel Use, INCO says.' *Toronto Star* (23 April 1992)

Evenson, Brad. 'Taking a Little Risk.' *Vancouver Sun* (25 August 1995)

'Greens Don't Want Fight, Mining Investors Told.' *Globe and Mail* (29 May 1992)

'Making the Public Aware.' *Mining in British Columbia* (Jan./Feb. 1990)

'Mining Company Sues B.C. Government.' *Globe and Mail* (13 July 1989)

'Modified Island Land-Use Plan Out This Week.' *Times-Colonist* (20 June 1994)

'NDP Gov't Has an Axe to Grind.' *Province* (21 June 1994)

'No Middle Ground.' *Province* (17 July 1994)

'Sudbury's Sense of Place Will Always Be Grounded in the Canadian Shield.' *Globe and Mail* (9 February 1994)

Toulin, Alan. 'Mining Coalition Presses Ottawa to Move on Regulations.' *Financial Post* (19 October 1995)

'Wilderness Brawl Looms.' *Calgary Herald* (15 October 1991)

Index